The New Metal Detecting The Hobby

Written by
Dick Stout

Edited by Mary Hand

Published By White's Electronics, Inc.
Sweet Home, Oregon

Dedication

To my wife Fay,
my two lovely daughters, Molly and Missy and to
Mr. & Mrs. Jesse Stout, affectionately known as Mom and Dad -
thanks for all your help and love over the years.
You are the most important treasures in my life!

Acknowledgements

Alan Holcombe, Marketing Manager for White's Electronics, for asking me to write the first edition of this book, and for always emphasizing integrity in the pastime and the business. He has been a good friend for many years, both to me and to the many thousands of detectorists across this country - thanks Alan!

Mary Hand, Marketing Assistant for White's Electronics, for the many hours she has put in organizing, editing and putting together all of my books over the past five years. She is the best!

White's Electronics, Sweet Home, Oregon, for allowing me to be part of their family, for making a fine product, and for donating so much time and money to promote, preserve and protect the treasure hunting pastime.

All of the many treasure hunters I've met over the years, both here and abroad. Your friendship and knowledge have made me not only a better treasure hunter, but a better person as well.

Photos by Fay Stout

ISBN #1-882279-01-8
Published by White's Electronics, Inc.
Sweet Home, Oregon

FOREWARD

Fasten your seat belt and hold on to your hat. Once you start reading beyond this page, you will be hard pressed to make yourself sit still long enough to finish reading this book in its entirety. Dick Stout is the consummate metal detecting artist with the unique ability to paint such compelling verbal images for you that your imagination will begin to race with excitement and anticipation.

It doesn't matter if you have never held a metal detector in your hands, or if you consider yourself a seasoned pro. In either case, I guarantee that you will not only learn something new, but be inspired in the process. You are about to be caught up in the spirit of that contagious enthusiasm that is Dick Stout. As always, he is eager and willing to share with you the wisdom, vast knowledge, and just plain good sense that he has acquired during his many years of involvement in the metal detecting arena.

This is your opportunity to learn from an experienced, seasoned detectorist, as well as one of the most modest and down-to-earth people you could hope to know. He doesn't presume to give you advice and tell you how you must do this or that, but rather relates the benefit of his own experience.

I invite you to enjoy Dick Stout here in these pages as I have had the good fortune to do in person. This book, like the camaraderie and the friendship we have shared metal detecting together over the years even when we found nothing, is irreplaceable. Good luck, but more importantly, HAVE A GREAT TIME!

Dan Hamilton - Pittstown, New Jersey

"Dan Hamilton is married to Mary, has two children, Colleen and Danny, lives in Pittstown, New Jersey, is a sales manager by trade, a seasoned detectorist via years of in-field experience and a sorely missed detecting partner and friend." - Dick Stout

Dick Stout with silver half dollar found along a canal towpath in eastern Pennsylvania.

The New Metal Detecting . . . the Hobby

Table of Contents

Table of Contents

Table of Contents

Table of Contents

Introduction

When I was asked to rewrite and update *Metal Detecting....The Hobby*, I was surprised and extremely flattered. I had always felt that it differed from other detecting books on the market in that it was based on my experiences in the field, it was easy to read and it was not filled with visions of grandeur and false promises. I guess those feelings were correct. *Metal Detecting The Hobby* went through three printings and I was encouraged enough to write two more books, *Coin Hunting In Depth* and *Where to Find Treasure.*

When I sat down to begin this rewrite, I read through the first edition again, cover to cover, and was surprised that even today, five years later, much of the material is still viable and useful. Yes, the computerized detector is now the norm, and visual target readouts are on even the least expensive detector, but interpreting their effectiveness and getting the most out of them is still up to the user. Like wise, the purpose of this update, like the first, is for it to serve as a *blueprint* for the beginning hobbyist, and a *refresher* course for the seasoned detectorist.

In this second edition you will certainly see changes. The use of the Internet and the world wide web has been a boon to the detectorist, and access to treasure has never been so close at hand. Where once we had to travel for information and research materials we now find it in the comfort of our own home. Wherever possible I've made mention of those web sites that have made a difference in my projects. Hopefully, you too will find them useful in your efforts.

To those of you buying this second edition of *Metal Detecting The Hobby*, I feel it's important to once again emphasize that I am not a professional treasure hunter, nor do I consider myself an expert in the metal detecting field. Like you, I'm a hard-working individual (*well, don't ask my wife okay!*), with a nine-to-five job and *secondly* a detectorist, treasure hunter, hobbyist, amateur historian, and simply someone who enjoys sharing his thoughts, ideas and adventures in the field. I have been metal detecting for 25 years, during which time I have found thousands of coins, some of them valuable . . . some not.

I have also found hundreds of rings and pieces of jewelry, again some of them valuable - some of them not. Add to these metallic finds other incidental treasures such as *hot wheel cars, keys, tokens, penknives, religious medals, a small cache of coins, and various other assorted collectibles*. What I've learned however, and what you will soon find out, is that it's not the good-to -bad ratio that makes this pastime enjoyable - it's the anticipation and the thrill of the search itself! It's the feeling you get when you hear that *solid* audio response, when you see that *positive* reading on your meter, and the excitement that you feel when you recover a silver half dollar six inches down.

One treasure that cannot be measured in terms of dollars and cents is the outdoor exercise associated with metal detecting. Being in the out-of-doors engrossed in your search is part of the pastime's reward, and I know of many who have taken up the hobby as the result of an illness or injury while others have found it a pleasurable endeavor as the result of retirement. The majority however participate because it's *plain ole fun*, and like the fisherman who never knows what his catch will be, so it is with the treasure hunter. How successful *you* become will depend upon how often *you* get out in the field, and how efficient *you* become with your metal detector, but rest assured you will have success the first day you use it!

I hope that after reading this book you will decide to join me and the thousands of other treasure hunters across the country. You will not simply enjoy metal detecting you will become addicted to it! It's a pastime that will remain with you throughout your life, and whether or not you find the big treasure that puts you on easy street, or that one key coin you need to complete your collection, you will certainly be richer for the experience!

Good hunting
Dick Stout/Rowlett, Texas/ January 1999

Chapter I - A Basic Primer

If you are new to the treasure hunting hobby, or thinking about getting involved, you surely have questions. Hopefully this chapter will give you the answers.

What Is A Metal Detector?

There is no need to bore you with technical explanations of how a metal detector is designed or manufactured, nor is it necessary to detail its inner workings and operational characteristics. Not everyone driving an automobile cares to take it apart or repair it, nor do those of us who watch movies on our VCR care to know how it functions. We derive pleasure from *"using them"*, and as a beginner this should be your approach to a metal detector. A plain and simple definition of a metal detector; **an electronic device that detects the presence of metal!**

All metal detectors have a **control box** to house the electronic components and batteries, a **searchcoil** that sends and receives signals, and a **stem** or **rod** that connects the two as one unit. When you scan the searchcoil over the ground a current is directed downward. When metal is detected, a return signal is sent to the control box producing an audible response through the speaker or headphones. In essence, a metal detector is both a *transmitter* and *receiver* all in one!

How Deep Will A Metal Detector Go?

The depth capability of any metal detector will vary depending upon the type of soil in the area being searched, the size of the searchcoil, the composition and size of the target, and of course user expertise. It's fair to say that in the beginning you should be able to detect a dime in the ground at six inches and a quarter at eight. As you continue to use a metal detector, and become familiar with its features, capabilities and characteristics, you will become more skillful, and get even greater depth.

Is A Detector Easy To Use?

Absolutely! Most all models offer *"turn on and go"* settings which allow you to set the controls quickly and find treasure the first day. Many metal detector functions that had to be operator-adjusted years ago are now handled automatically by the instrument's internal circuitry and/or microprocessor. If you can flip a switch, turn a knob or press a pad, you can start metal detecting.

Along with coins you can expect to find an array of metallic items - many old and quite collectible.

What Can I Expect To Find?

You can expect to find anything metallic in composition. You will find items made of iron, copper, brass, aluminum, silver and gold. Your goal of course is to find coins, rings, jewelry, relics and other collectibles, and with the advanced capabilities of today's metal detectors this can be accomplished quite easily.

The rings pictured above were found during one year of searching.

Are There Really That Many Coins To Be Found?

This may surprise you - there's more money in the ground than in circulation. The number of coins you find will depend upon how often you use your metal detector, and how much time you put in each time out. It's not unusual for a coinhunter to return home with over one hundred coins in one outing. While the total amount of these finds may not be much, the grand total over a year's time is certainly worth the effort. Many of my detecting friends average from $1,000 to $3,000 a year, and they are *"part- time"* hobbyists.

The quantity and quality of your finds will depend upon the area you are hunting, its age, and whether or not it was used by hundreds of people or a few. The best part is, as coins are found they continue to be lost. An ongoing replenishment cycle!

What Kind Of Coins Will I Find?

Here again the area you hunt will determine whether you find newer coins, older coins, or a combination of both. Dimes, quarters and halves minted prior to 1964 are 90% silver. With the current price of silver at around $4 dollars an ounce, a common date silver coin can be worth three to four times its face value. Others, because of their date, scarcity, and low mintages - much more!

Coins minted after 1964 are referred to as *"clad"* and tarnish quickly in the ground. A silver coin will look just as shiny as the day it was lost. You may also be lucky enough to find a gold coin or two, and just one can pay for your metal detector. Gold coins were legal tender years ago, and were lost from time to time, but not with the regularity of other lesser valued coins.

Note the tarnished clad coins on the left, and the shiny pre-1964 silver ones on the right.

How Much Does A Metal Detector Cost?

Metal detectors range in price from $100 to $1,000. Higher priced models offer extra refinements and somewhat better depth, but are not necessary in the beginning. A metal detector in the $200 to

$400 range will serve you quite well, and will start you on your way to being a successful treasure hunter.

Where Can I Purchase A Metal Detector?

Write or call metal detector manufacturers, and request a copy of their catalog, as well as the names and addresses of dealers in your area. This can also be achieved by visiting the web site www.treasurenet.com. Lastly, look through your telephone company's yellow pages under *"Metal Detectors"* for additional assistance.

Purchasing a metal detector through a factory authorized dealer will allow you to see the various models, learn about their features, and actually try them. A local dealer will also answer questions that come up later, and can handle repair problems should they develop. For the most part, a metal detector will last you a lifetime.

What Should I Look For When Shopping For A Metal Detector?

Make an effort to purchase a detector that is comfortable for you to sweep. Long hours of scanning can sometimes cause arm fatigue and loss of concentration. A model in the four to five pound category is acceptable, and fairly standard today. Most manufacturers also offer hipmount convertability. This means you have the option of removing the control box from the stem, and wear it on your waist or chest with a strap. This option significantly reduces the overall weight and strain on your arm.

Next be sure that any detector model you consider has both **Discrimination** and **All-Metal Modes** (this should not be difficult since most all models do today). Discrimination circuitry allows for trash rejection and the acceptance of better targets, while the All-Metal Mode responds to all metallic items. How, where, and when to use each mode will become clearer as you continue reading.

Most metal detectors manufactured today come standard with a *searchcoil seven to nine inches in diameter.* This size is best for most of your endeavors. Larger and smaller coils are also available for specialized applications, and will be discussed later.

Target Identification Displays are not necessary when starting out, but it's almost impossible today to find a detector without one. A target ID display will tell you when you've found a penny, nickel, or dime, and also identify many of your trash items. This saves you a great deal of wasted effort in the field. It's important to understand that Target ID displays are not foolproof, and are *probable* target identifiers only!

Your first metal detector will surely have additional features or refinements, but I will not attempt to guess what they might be, nor will I take the time now to describe them all. If you purchase a detector with *discrimination,* an *all-Metal mode,* and *a seven to nine inch searchcoil* you are ready to find treasure. Other features will be bonuses or *"icing on the cake".* Be sure the dealer explains each feature to your satisfaction.

At this time if you feel you must know more about the various features and functions, refer to the glossary at the end of this book.

Is A Metal Detector All That I Need?

Besides your metal detector you will need to purchase four additional items from your dealer. The first is a set of *headphones.* No matter how expensive or sophisticated your metal detector is, you must use headphones if you are to be successful. Yes, your detector does offer an external speaker, but using it will not be to your advantage. At best you will only hear the large and shallow items, and in the process miss many of the smaller and deeper targets. Using headphones will prolong battery life, and allow you to concentrate on each signal by reducing outside noise such as traffic, crowds and the wind.

Be sure the headphones you purchase are comfortable to wear, and have individual volume controls for both ears. You may not think you have a hearing problem, but you will quickly discover that one ear is less sensitive or lacking, and having individual controls will allow you to adjust accordingly. A good rule of thumb is to set your detector's volume as loud as possible, then adjust the volume on your headphones.

Using headphones will prolong battery life, and allow you to concentrate on each detected target, especially the smaller deeper ones.

How much should you spend on headphones? My advice is not to buy the most expensive or the least expensive. Generally, headphones in the $40 to $100 range will get the job done! My favorite - White's KW Pro.

The next item you will need is a *digging tool* for recovering your finds. Over the years I have used all types of implements, from screwdrivers to knives, to trowels and pointed probes. The most popular digging tool is the garden type trowel, but allow your dealer to recommend one that is durable and works well in your area. You will encounter all types of soil while metal detecting. Having a decent recovery tool is important.

Next is a *pouch or apron* to store your finds. A carpenter's apron with two pockets works allowing one compartment for good finds and one for trash. Always pick up all trash, and dispose of it in a trash receptacle. It's part of the hobby's code of ethics, and frankly common sense since you may return to the same site later and end up digging it again.

The last accessory (which shouldn't cost you a cent) is a *"dropcloth"*. A dropcloth is a piece of cloth or plastic, approximately 18" by 18", and will be used to temporarily hold any soil you may remove when recovering a target. This will be explained in Chapter II.

A good set of headphones, a durable digging tool, a probe and an apron are all you need to get started.

Where Can I Use A Metal Detector?

The ideal place to metal detect is where people congregated in large numbers, and for long periods of time. The more an area has been used, the more treasure it will contain, and the older the site, the better your finds. As you continue metal detecting you will read, talk to others and research - seeking those sites that offer you the best return for your time and effort. As your interest grows, so will the quality and quantity of your finds.

At first try detecting schoolyards, parks, beaches and picnic areas. These sites always give up a few coins, and are ideal for learning more about your new metal detector. Later on in Chapter 3 I'll discuss additional sites that are also good for the detectorist.

Schoolyards and parks will always give up a few coins, and in the process allow you to better understand your detector's functions and capabilities.

Chapter II - Metal Detector Fundamentals

Assuming you've decided to become a treasure hunter, let's go over a few metal detecting fundamentals. There's nothing complicated about operating a metal detector, but many in their haste to get started, develop a few bad habits. The purpose of this chapter is to make you aware of them, and by doing so increase your odds of success.

The Owner's Manual

After purchasing a metal detector be sure to read the owner's manual from cover to cover. After doing this read it again! This manual is written with the beginner in mind, but rushing through it may cause problems later on. Read it thoroughly and keep it with you at all times for reference.

Your Own Practice Area

To better understand your metal detector, its functions and operating characteristics, lay out a practice area in your backyard. If possible it should be a minimum of twenty feet by five feet in size. Bury coins, a jewelry item or two, and a few trash items (nails, pulltabs and foil). Use different denominations of coins, and while you are not interested in finding trash items, be sure to include them so you will understand your detector's reaction to them, and ability to reject them. Write down the location and depth of each item you bury (you will remember initially but probably forget later).

There's no set blueprint for your target area, however, be realistic and do not bury any item excessively deep! Three to four inches is a good practice depth. After you have used the test area for a while, you can redo it moving items to different depths or substituting other metallic targets. Familiarity with your detector, time and practice will dictate this.

Once you've completed your practice area, begin scanning over the items you buried. Listen closely to how your detector responds, and if you have a target ID display, note their readouts especially the deeper ones. They may not offer enough of a signal to trigger the correct readout. Remember this when you venture out in the field.

Experiment with your discrimination control, and learn where the trash items are eliminated. Know each metallic rejection setting (your discrimination controls may already be labeled, but every detector is slightly different, and rejection points may vary slightly from the factory settings). I have found the "White's Target Stick" extremely useful for this as well as for learning target readouts and audio responses.

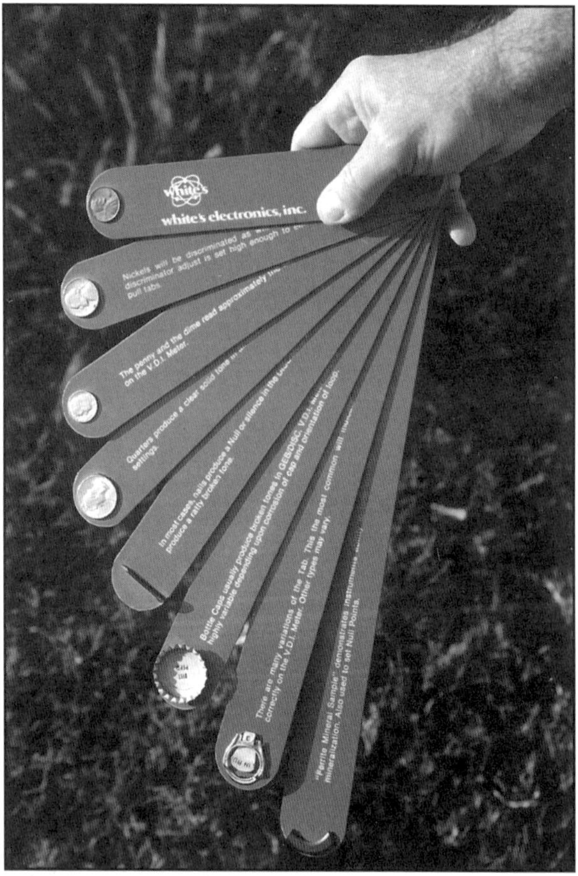

White's Target Stick is a useful tool for learning precise discrimination settings, and for audio/visual target recognition.

Proper Scanning

The process of scanning the searchcoil over the ground is probably the single most important factor in being a successful detectorist. In case you don't believe this, observe other detectorists as they search an area. Notice how quickly they move, how much ground they cover, and how lackadaisical they are as they go about this task. Chances are they're thinking about everything "under the sun" **except** covering the area thoroughly. Understand that this is normal, and not an easy habit to break. Scanning with a detector becomes very mechanical, and like sweeping the floor or raking leaves, the mind tends to wander and attention to the task at hand lessens.

When searching any site, scan very slowly. Because your searchcoil produces a cone shaped pattern in the ground it's important that you overlap each sweep by *at least one third to one half* for adequate coverage. One way to make this happen is to shorten your stem or rod so that your searchcoil is literally at your feet. This will allow you to visually make sure you are overlapping, and mechanically it will cause you to slow down (try moving fast with the coil at your feet).

Another common mistake many detectorists make is lifting or *"dishing"* the searchcoil at the end of each sweep. Again this is not a habit that's easy to overcome. The simple act of scanning the coil from side to side makes lifting seem natural. Keeping the coil parallel to the ground throughout the entire scanning arc necessitates a great deal of concentration. Assuming your detector may have a maximum depth capability of eight inches, lifting your coil on each sweep can cost you at least three to four inches in depth and perhaps four feet in ground coverage. **Keep the coil parallel to the ground at all times.**

Last but not least, keep your searchcoil off the surface of the ground. While coming in contact with the ground will not cause damage to the coil, it will cause *"backreading"* to occur, meaning that some trash items may respond as good targets. To better understand backreading, turn on your detector and set your discrimination control to reject pulltabs. Next take a pulltab and move it back and forth beneath your searchcoil, keeping it at least an inch away. Note that it does **not** respond audibly. Now *rub* the pulltab across the surface of the coil and see what happens. The positive response that you hear is backreading. Keeping your coil slightly elevated will eliminate this problem, and the time you spend digging worthless trash items.

Audio Threshold

Your metal detector may offer both silent or threshold operation. What this means is you can choose between hearing a faint constant threshold hum, or hearing absolutely nothing but your targets as they respond. Silent operation has become popular over the past few years, but if you have a choice choose the minimum threshold.. Utilizing the audio threshold will enable you to hear faint responses indicative of deeper and (usually) older coins. These barely audible signals often are not strong enough to push the audio from silent to audible. With an audio threshold you will hear some change when these deeper targets are detected. There are some I'm sure who would argue this theory with me, but my experience has taught me otherwise.

Target Responses

When searching an area, you will receive many different audio responses. Over time you will learn which are good targets and which are not. Typically a good target will produce a solid audio response that can be repeated no matter which direction you move your coil over the target. On the other hand, trash items tend to produce scratchy or broken responses. You might also hear a solid response in one sweep direction, but a scratchy one coming back. This can happen when a good target is close to a bad one, causing the discrimination circuitry to become confused. To solve this problem try scanning the target from different directions, and attempt to isolate the good one from the bad. Hopefully, you will be able to produce a solid response on both sweeps, and recover the better target.

One particular trash item that frequently causes problems is the aluminum can! Most are found within the first four inches of the surface, and because of their conductive metallic composition and size, they respond like good targets. To determine if the target is an aluminum can, switch to your all metal or pinpoint mode, and lift your searchcoil about a foot above the target. An aluminum can will continue to give a loud and pronounced signal, whereas a coin sized target will diminish or lessen in intensity. Utilizing this very simple procedure will save you a great deal of wasted digging, especially in trashy areas.

Interpreting the various audio responses will become easier after a little practice in the field. Be patient in the beginning and realize that you will dig some trash (in fact you may even **want** to dig every signal you receive to better understand how your detector reacts to various items). You will soon become a pro at telling the difference between trash and treasure.

Discrimination

Probably the most useful feature on your metal detector is *discrimination.* If it were not for discrimination, your detector would sound off over every metallic target in the ground, causing you to dig every signal to see if it's good or not. Discrimination circuitry works on metallic conductivity. It offers you the opportunity to reject various metals you do not want to detect, allowing those you do to sound off audibly. Sounds simple doesn't it? Well unfortunately there's a little more to it than that!

As you turn your discrimination control clockwise you will begin rejecting various metals. Fortunately, most of the least desirable (iron and foil, etc.) are eliminated first making it possible to accept the better metals like copper and silver which are at the upper end of the scale. The problem area, and what confuses most detector users, is the *middle range*.

Because gold and aluminum, which fall in this mid-range, are so close in their conductivity it's virtually impossible to reject one without rejecting the other. This means that some gold rings and gold coins will be eliminated if you set your disc control to reject aluminum. That being the case, what do you do when hunting a trashy area loaded with aluminum pulltabs? Unfortunately there's no simple answer! It's a dilemma all detectorists face, and the decision where to set your disc control is not an easy one. I have learned to deal with the problem by analyzing the area I am hunting, and what it might hold in the way of treasure. If I feel strongly that it has the potential to perhaps produce a gold ring or coin, I set my discrimination lower accepting the fact that I will likely dig more trash. In other words, the rewards are worth the extra effort. On the other hand, if the site is extremely trashy and doesn't offer a great possibility of gold coins or rings, I turn my disc control up, reject the pulltabs, and still recover most of the coins present (excluding nickels, which fall even lower on the conductivity scale).

Experiment with discrimination settings and various metals. Learn where each is rejected. Knowing how much discrimination to use, and when to use it is very important. Setting this control too high can cause you to miss valuable targets, and lessen your depth capabilities as well.

Target ID Displays

Target ID displays are valuable aids, but here again it's important to understand their purpose and limitations. When target ID displays were first introduced about ten years ago, hobbyists couldn't wait to purchase a metal detector with one. Imagine being able to tell what the target was before digging it? I purchased a top-of-the-line model, found the metered ID fascinating, and after using it for a while became mesmerized by it. Unintentionally and unknowingly I became a looker instead of a listener! When my better finds began decreasing, I realized what had happened, and went back to concentrating on the audio using the ID display as a *secondary* check only!

Target ID displays can tell you when you've found a penny, dime, quarter or even an aluminum pulltab. What you must remember is that they are advertised as **PROBABLE** target identifiers only! They can be fooled. If you look and do not listen, you will surely leave good finds behind. ID displays are at best 80% accurate, and should *only be used as a secondary analysis of the target.* Like the audio threshold problem, if a coin is buried deep or on its edge in the ground, the signal produced may not be strong enough to push the display readout into the correct ID area causing a false reading.

Target ID displays are becoming much more sophisticated today, and a few manufacturers offer LCD (liquid crystal displays), numerical target readings, and identifiable graph patterns. By comparing each of these we are beginning to see better target patterns and definitions making identification more accurate.

Proper Target Recovery

You now understand how to detect various targets in the ground, but what about actually recovering them? How will you accomplish this? Knowing how to retrieve your finds is important for a number of reasons. First and foremost, you want to leave any area you detect exactly as you found it. Next you want to recover the item in the quickest way possible without causing damage, and last you want to be accepted by your metal detecting peers. Not utilizing proper recovery techniques affects everyone who participates, and frequently gives the pastime a bad reputation.

Pinpointing

Once you have decided that a target is worth recovering, you will need to pinpoint its location in the ground. If your metal detector has a pinpoint mode (all metal) use it! When the audio response is at its peak you have centered on the target, and it should be beneath the center of your searchcoil. If you do not have a pinpoint mode, move the coil back and forth over the target making an imaginary "X". This will give you the approximate location of the item. Keep eye contact on the spot and prepare to recover the target.

Properly recovering your treasure is just as important as finding it.

Plugging

After pinpointing the location of your target insert your knife or garden trowel in and around it, and remove a cone shaped section of sod. Be sure the angle and depth is adequate to encompass the roots of the grass so that after replacing the section it will continue to live. Hopefully you will be successful, and your target will be in the section of sod you remove. To be sure, move it back and forth over your searchcoil. If the target is in the plug, the detector will respond. If it does not, scan the hole and attempt to locate the target again. You may need to remove additional soil or enlarge the hole. If so, use your dropcloth to hold the soil. After recovering your target, pour the loose soil back into the hole and replace the plug. Tap it firmly with your foot before resuming your search. If you have been successful in your recovery, the area will look virtually untouched.

While removing a plug is fast and easy, there are times when it is not appropriate. For instance, if you are detecting during the hot dry months of late summer or during a period of prolonged drought, a plug may wither and die leaving unsightly brown patches. If you face these conditions, it's best to return at a later time when you are certain that you will not damage the area.

If your pocketbook permits, I would also recommend you purchase White's 120mm Bullseye! This miniature handheld probe will tell you exactly where in the plug or hole your metallic target is. Simply hold the button in - move and point. It will audibly sound off when you locate the target. These small handheld probes have been popular in the UK for sometime, and it's nice to see them available in the US.

The Bullseye is an easy to use, and effective tool for pinpointing my targets.

Probing

Another excellent method of retrieval is probing! A probe is an *"ice pick"* type of tool approximately eight inches in length with a blunted or annodized point (to avoid scratching a valuable find). Your detector shop can recommend one that is suitable. Probing may take a little time to master, but once learned you will be ready to detect even the most manicured of areas. It's a much more precise way to retrieve your finds, and should always be used when hunting well-kept lawn areas.

Once you have pinpointed your target insert your probe and actually make contact with it. At first this may seem difficult, but eventually you will be successful! Once you locate the target, make a vertical slice in the turf, keeping the point of the probe stationary. Now do the same thing horizontally. This will leave you with a cross or "X" in the turf, and your target in the middle. Next insert your probe under the target, and gently pry it to the surface being careful not to scratch it in the process. After you've recovered your find push the folds of the turf together, step on the spot and continue your search.

Which retrieval method you use is up to you, but it may also be dictated by the soil, weather, and the area you are detecting. With time and practice you will become adept and comfortable doing either. Generally speaking, when you are detecting a rough or unkept area, it's acceptable to remove a plug. The better maintained the area or the drier the ground - probe!

Chapter III - In The Field

When you first venture out with your metal detector you will probably be searching schoolyards, parks and other similiar places that quickly come to mind.. You'll be anxious to find that first coin, and to see if your detector works. Trust me - the detector does indeed work, and you will find money, but be patient during this startup phase. Trial and error will be your teachers, and soon turn you into a detecting pro.

The First Time

For some reason most detector users feel a little self-conscious the first time they venture out. I am not sure why, but I suspect it's because they think (1) the metal detector really doesn't work, and (2) their friends and neighbors will laugh at them. Whatever the reasons this uncomfortable feeling soon passes, and you find yourself looking forward to every opportunity you have to go detecting. Every item you uncover will be cause for excitement, and the desire for older more valuable finds will naturally develop as you continue hunting. Have a positive attitude, remember the fundamentals and last but not least, **have fun!**

Searching Private Property

Over the months ahead you will be detecting both public and private lands, and it's important to understand your obligations with regards to both. Public areas like parks, schools and beaches almost always allow metal detecting, but you should always check for posted signs or restrictions that may state otherwise. When detecting private property **you must obtain permission from the landowner!** If not, you are trespassing!

When you want to search private property decide on how you will approach the owners. You may want to knock on their door and ask in person (the best method), phone or write to them. I have found they all work well, but I much prefer meeting face-to-face with each landowner. No matter which approach you decide on be forthright about your purpose, be polite and always appreciate the answer no matter what.

If you are the nervous type or somewhat shy, writing a letter may be the answer. Begin by introducing yourself, and explain why you are writing. Tell the property owner where you live, where you work, why you would like to detect his property, and be sure to emphasize that you will not damage their property in any way. You might also mention that you will sign a waiver alleviating them from any liability. Include a self-addressed, stamped envelope for their reply, or suggest that you will phone in a few days for an answer. Writing will frequently work in your favor since it does not necessitate an immediate answer, and gives the property owner time to consider your request. Providing personal data also allows them time to check on your character and standing in the community if they so desire.

If you seek permission in person, do it in the best possible fashion. Dress neatly, be polite and whatever you do, do not show up at the property owner's door with a metal detector in one hand and a digging tool in the other. This is presumptuous, and makes for a very uncomfortable situation. Introduce yourself, explain why you are there and again emphasize your expertise in recovering targets. If there's any reluctance, allow them to think about your request, and contact them at a later date for their decision. Some hesitation is typical. Many do not understand the hobby, are not sure what to expect, and as a result find it difficult to give a quick answer.

Remember too that liability is important. It used to be that an individual's word was enough, but with the number of injury related lawsuits in the courts today many landowners are overly cautious. Draft a short disclaimer that you can take with you. Sign it and leave it with the property owner so that he will feel more comfortable making his decision.

There's really no secret to accessing private property. It's *common sense!* Approach every property owner as **you** would wish to be approached under the same circumstances. You only get one chance to make a good first impression so make it count. Remember, patience is very much a virtue! I have had property owners turn me down, only to call at a later date giving me permission, and frankly, a *"no"* answer is always a possibility, but consider it *"status quo"*. You couldn't search the site before you asked, and you cannot search it now. *Nothing ventured, nothing gained!*

Over the years I've found that gaining access to private property is usually easy. The odds are certainly in your favor, but your success will be based on your approach, your presentation, and your appreciation for each situation.

Where To Search

There are literally hundreds of places to metal detect, and I am sure you already have a few in mind. As you continue to search various sites, you will constantly be looking for others in order to find even older and more valuable coins. Let's talk a little about the sites that immediately come to mind, and what they have to offer.

At school you will find your first treasure and more than likely a few other surprises.

Schools

There's probably no better place to begin your metal detecting hobby than the local school yard. Here you will find your first coin, and most likely a few other small treasures. More important, you will become familiar with your metal detector under typical conditions. Children play and lose things making the schoolyard not only ideal for the beginner but for the experienced Th'er. It's usually close by, sure to provide a multitude of signals, and just maybe something of interest and value.

When you begin searching the school area remember to scan slowly, keep your searchcoil parallel to the ground, and overlap your sweeps by at least one third. Analyze all of your target responses, and make a mental note of what you recover. Don't be in a hurry. This first outing is as much a learning experience as a chance to find a few coins.

This old one room shool in rural New Jersey was completely overgrown in the summer, but very accessible and productive in the fall and winter months.

If the school offers such, search around the playground equipment (swings, seesaws, monkey bars and sliding boards). Ballfields and lunch areas are also ideal for finding coins as are slopes and hills. Searching these areas will certainly keep you busy, but don't stop there. The perimeters of the schoolyard (and for that matter any site) can be productive, and are often overlooked by many hobbyists. I am not sure why but I suspect it may have to do with the area being a little larger years ago, and those taking care of it became lazy allowing it to shrink in size.

You will not become a millionaire at a school, but you will find coins each time you visit. You may also be fortunate enough to live near an older school (used prior to 1964), and find a silver coin or two. Whatever the outcome, you will fare well by going to school!

Parks

If your town or city has a park spend some time detecting it. Most have been in existence a long time, and frequented by many people who surely lost valuables and coins. City parks are almost always public property, but you would be **wise to check first.**

When you arrive at the park look around and decide where you might begin your search. What does the park have to offer? Is there a bandshell, a ballfield, picnic area or play area? Is there one area that seems to attract a larger concentration of people? Observe the activities taking place, and decide where to begin your search.

Talk to people, read and find out all you can about the park(s) in your area. Find out how long they've been in use. Again the older the park the older the coins! Some will have changed in appearance over the years, and finding old photos will aid you in your efforts to locate the better "hot spots".

A reminder.... when people visited a park they usually had to negotiate the grassy strip between the street and the sidewalk. This narrow area is overlooked by most hobbyists yet often very productive. It's also *public* property in most communities.

Occasionally you will find a park that is unkept, but most are well maintained so remember to **probe your targets, use a drop cloth,**

likewise never interfere with activities that might be taking place. You always have the early morning hours to search if you are faced with this scenario.

The community park is a good place to make use of your dropcloth.

Churches

Churches are a favorite of mine, especially the very old ones, but I must remind you... they are **NOT** public property, and you will need permission to search their grounds. Most pastors or priests will allow you to detect if you approach them in an appropriate manner, and when given permission be sure to search at a time when you will not interfere with church services or activities.

Years ago it was very common for rural churches to hold church suppers. They were sometimes held outdoors during the summer months, and close by in a wooded grove. Harvest Home Suppers, as they were sometimes called, were very popular events, often being the only social get-togethers available in smaller communities. Make an effort to locate this site. These old groves are excellent for detecting, and often still nothing more than wooded areas, very accessible and frequently void of modern day trash. I was very fortunate to search a few that offered nothing later than turn of the century coins!

Older churches, such as this one in Rowlett, Texas, are favorites of mine.

HOMESITES

An old homesite is capable of producing old coins, a few relics, and perhaps a small cache of money. It might consist of one building, many buildings or nothing at all but a plot of ground. Old homesites exist in every area of the country. If you need help finding them, two of the most reliable sources I know are the *outdoorsman/hunter* and the *farmer.* Their interests and chores take them off the main road and into many areas where foundations and remnants of homesites are hidden. They will not generally offer this information because they assume it isn't of any interest to anyone. As a detectorist they are of great interest to you. Ask for their help.

You will also find old homesites by being more *observant.* Because of your new hobby, you will begin to take notice of signs of habitation that have been around for some time (remember prior to purchasing a metal detector you didn't have a reason to notice). Look closely at any open field you may pass while driving. If you see a cluster of trees in the middle it might be they're camouflaging an old home or homesite foundation. Trees have always provided beautification and shade for dwellings, and are often all that is left to indicate their existence. Other clues - plants or vegetation that does not blend in with the surrounding area. Foreign or unusual plantings are easy to spot, and an indication that someone earlier inhabited the area and made an attempt to beautify it. Another clue that often escapes notice is the camouflaged driveway or lane that appears to go nowhere. All roads were constructed for a purpose, and the old and forgotten ones can often lead to treasure. Yes, it may be they are only an access to a field for planting and plowing, but it's also possible they will lead you to an old homesite. *It only takes a few minutes to investigate!*

This old farmstead was littered with trash and difficult to hunt, but it was well worth the time and effort.

When detecting a homesite, be sure to search all of the buildings involved for *money caches.* Caches have been found hidden in doorsills, windowsills, under floorboards, in outhouses and under stone walkways. They have been found under fence posts, in stone fireplaces, in old tin cans on barn beams, in old pieces of farm machinery and under porches. Remember too that they are probably stored in a container of some sort to protect them from the elements so utilize your all-metal mode and check every metallic signal you receive. Yes, it will take time but it's time well spent considering the potential rewards!

The old homesite offers coins, relics, old bottles and just possibly a long-forgotten cache!

Depending upon age and past use, homesites can be difficult and tedious to hunt. Many are littered with small pieces of iron and junk, but in-between is a coin or two worth your time and patience. The hard part is knowing just how much discrimination to use. Because of the overriding possibility of finding a cache, I recommend setting your disc control quite low. As you recover trash, place it in one designated area. Eventually you will open up the area you are searching, and in the process find the better items that are present. I also recommend that you keep homesites in mind for *"continual"* searching. You may not find much the first time or two, but persistent searching will almost always produce something of value.

Finally, if you are really ambitious you might try to locate the homesite's trash pit in quest of old and valuable bottles which bring good premiums when sold to collectors and antique dealers. Trash pits or dumpsites are easily found by locating the area that produces an unusually large audio response, and while it may be hard work, delicate digging with a shovel will almost always produce an old bottle or two of value.

Picnic Groves

I purposely use the word *"grove"* because I am referring to **older** picnic areas that were the social centers of many small communities. Now forgotten, they sit waiting for someone to recover their treasures. Television and radio were not options for those at the turn of the century, and entertainment was getting together with friends and family at the local grove. Ask the old-timers about them and ask where they were located. If you search one you'll start searching immediately for others in the area.

How to find picnic groves? This road sign might lead the way.

How can you recognize an old grove area? What should you look for? *Large trees to provide shade, very little ground vegetation, frequently a small creek or stream to use for wading or swimming and a lane or entrance that allowed access.* Generally speaking, picnic groves were never far from town, and many times a street or road name will often lead you to them (Elm Grove, Oak Grove are good examples). Old groves rate a 10+ with me, and account for many of my better finds. Once you spend a quiet day in the woods finding silver coins over 100 years old, you will truly know what metal detecting is all about.

Swimming Areas/Beaches

Most everyone will tell you all they know about metal detecting is it's something people do on the beach. While this is not entirely true, some hobbyists do **only** use their detectors for beachhunting. The reason? It's relaxing, healthy and one valuable ring or piece of jewelry can pay for a metal detector, or a family's summer vacation! Detecting the beach will routinely pay off with a handful of coins, but the added bonus of gold and silver jewelry makes it even more appealing. Swimmers and sunbathers do not fully realize the effects of water on the hands and fingers, **treasure hunters do!**

Ocean tides will play a big part in how much treasure you will find on a beach. If you doubt this, visit one after a large storm or hurricane. You'll see scores of detector users searching as though there's no tomorrow. These storms produce above average tides that carry away large amounts of sand allowing deep buried treasures to be exposed, and in the process *bring in* treasures from offshore. ***Strange as it may seem, bad weather for detectorists living near the coast means "good hunting"!***

Remember your goals when metal detecting a beach and avoid using too much discrimination. Rejecting aluminum pulltabs will mean losing gold rings and jewelry, your main objective! Yes, you will be digging pulltabs but the rewards are worth it, and digging on the beach is not nearly as time consuming as on land.

I would be amiss if I didn't mention community *swimming holes and lakes,* inland hot spots that should always be sought out. A community swimming hole may not have been more than a deep

Detecting the beach is not only profitable, but very relaxing.

spot or bend in the creek. But whatever it was or wherever it was it usually attracted many people during the hot days of summer, and can provide you with good detecting. Inland lakes also offer recreational areas similiar to coastal beaches for swimmers, boaters and picnickers and can provide the detectorist with similar treasure.

The ropes hanging from this large tree led me to this old swimming hole in a meadow.

Athletic Fields

Athletic fields of all types, hold treasure, and as I said before, the older the area the better the finds. Whether the field was used for baseball, football or soccer, it attracted crowds and crowds mean lost coins and valuables.

An old high school football field in New Jersey became a favorite of mine accounting for literally hundreds of silver coins. After hunting it over 150 times, I still managed to find a few on each visit. One day in particular I decided to work the outer perimeter of the field to see if I might scare up a few coins I had missed. Suddenly in an area approximately twenty by twenty feet, I began finding a number of coins most all of them silver. I was excited but also curious why and how this small plot of ground was giving up so much. It was months later while reading an old high school yearbook that I found the answer. It was the entrance area to the field where admission fees were collected. Evidently the hustle and bustle to enter, cold

Athletic fields continue to attract participants, spectators, and detectorists.

hands, and the exchange of money accounted for my good fortune - a hotspot within a hotspot!

Most athletic fields have been hunted by metal detecting hobbyists, but you can never assume that all the better treasures were found. Searching around bleacher seats will be the best place to start. If the field is quite old and no longer has bleachers, do a quick search of the area looking for the greatest concentration of signals. This will usually indicate the bleachers original location, and a good place to begin your search. If bleacher seating does exist and your detector starts sounding off because of the metal supports, decrease your sensitivity. Small coils (discussed later in this book) also work well in this instance.

Finally never assume that the athletic field of today was the athletic field of years ago. It might be that the ballfield of the turn of the century is nothing more than an overgrown field today. Read all you can about the schools in the area, and about community activites. **Time changes many things, and if you are to be successful you must always remember that what you see is not necessarily what you'll get!**

Carnival Grounds/Circus Sites

There aren't many communities or cities across the country that haven't held carnivals or had a circus come to town, and more often than not they were staged in the same area or field. If this is the case you will likely find a lot of litter and trash, but don't give up! I have hunted carnival and state fair sites that were almost solid bottlecaps and pulltabs, but because I knew there were also old and valuable coins present I perservered and came away with quite a few nice finds. You can overcome this problem by using extreme discrimination settings or by working a very small area cleaning out the trash that is present, and in the process finding the better coins. The latter option is back breaking work but worth the effort since most other detectorists usually opt to move on to another area.

Scout Camps

Most Boy Scout and Girl Scout camps have been around for some time, and can offer excellent detecting. Check with your local scouting leaders or groups to find where they are located, and how to obtain permission to hunt them. Accessing scout camps is usually not difficult, especially in the off season of early spring or late fall. They can provide acres of lost coins and valuables and a site to return to again and again.

In 1982 I gained access to a very old Boy Scout camp in New Jersey in exchange for giving a class or two on metal detecting. I enjoyed these presentations, and in the process hunted the area for six months coming up with many silver coins. Why not offer a similiar plan? It's painless, it's fun, and you'll enjoy some great treasure hunting.

This old lakeside pavilion was all that remained of an old Boy Scout Camp.

Ghost Towns

No matter where you live there are ghost towns. One thinks of finding them only in the Old West, but that's not the case...they are everywhere! You may doubt this, especially if you've lived in your area for some time. Well, I challenge you to visit your local historical society or library and find out.

Early settlements and villages were usually centered around a particular industry or trade, and their growth was dependent upon its success. When industries failed, lack of work, and the inability to make a living sealed the fate of these communities many disappearing without a trace that they even existed. While researching your local history you will likely come upon the mention of just such a town or village.

Forget the stereotypical picture of a ghost town. They vary in appearance, and often only the hint of a building or a foundation will be your clue all too often hidden by overgrowth and vegetation. I also hunted one ghost town that was nothing more than a very large field, with absoluely no indication or clue that it was once a thriving community. With research you can find them, and recover all the treasures they have to offer.

Other Good Areas

You will continue to look for new places to metal detect, and that's part of the fun. Here are a few other areas you may want to consider and seek out:

Playgrounds, picnic areas, campgrounds, rest stops, sidewalk strips, carnival sites, circus grounds, rodeo arenas, rural mail box areas, old tavern sites, revival meeting sites, military bases, swimming areas, beaches sledding areas (*hills*), corner lots, towpaths, flea markets, train depots, amusement parks, family reunion areas, groves, campgrounds, old fort sites, rural dance areas, and band shells,

and. . . .

town squares, construction sites, old farms, fishing holes, mining camps, general stores, trailer parks, hiking paths, orchards, outhouse sites, battlesites, arena sites, backyards, racetracks, fence rows, boat ramps, resort areas, ski areas, old movie drive-ins, motels, roadside stands, college campuses, under boardwalks, market areas, telephone booths and ????

Summary

A great deal of the fun I've had detecting over the years was in researching and finding new sites. New unexplored areas are exciting, and being the first to detect them can be very profitable. If you are persistent you can be the first, but if by chance you aren't don't be discouraged. No site is ever cleaned out! No matter what others tell you, there will always be treasure left to find. Each and every site has a way of replenishing itself. It's one of those mysteries of the hobby that cannot be explained. You can continually hunt an area, day after day, year after year, and **always** find something.

When someone tells me they've hunted out a site or found all there was to be found, I make a concerted effort to detect the area! What that means to me is they have gone to the trouble of cleaning out the trash, and what's left is old, a little deeper, and requires more effort than they care to put forth. By maximizing my detector's functions and going slow, I am confident I can find the best the area has to offer!

The list of sites I have mentioned in this chapter is by no means complete. With a little thought you will come up with others, and always have success if you remember *where people congregated treasures were lost. The more people, the more treasure. The older the area, the more valuable the treasure!*

Chapter IV
Research The Key To Success

As you continue to metal detect you will not only be looking for new areas to search, you will become increasingly anxious to find older and more valuable treasures. If you are to satisfy this anxiety you must take the time to read and research. When one uses a metal detector there will certainly be a few discoveries made as the result of random and chance searching, but the most frequent and most rewarding are the result of diligent and ongoing research efforts.

Each new site you discover will spur you on to the next level. Another site, only better, another find, only older and more valuable. They are all around you, and whether or not you're the first to find them will be up to you. Those that know me personally, or have read my articles know how I feel about research. It's the difference between success and failure! The difference between average and expert!

Why Research?

Research because it will not only lead you to treasure, but act as a catalyst and add to your overall enjoyment of the hobby. As you immerse yourself in this part of the pastime you will sometimes become frustrated, out of clues, and no where to go. Then suddenly you'll happen upon a bit of information that adds one more piece to the puzzle, and takes you one step closer to your goal. Not all of your projects will pan out, but those that do will keep you going.

When researching, be sure to make use of all available resources to find the treasures you seek. Read local history texts, talk to people, and study maps and old plattes. Go to the Internet, surf the web, putting into play any and all keywords you can dream up. Write government agencies and individuals, and last but not least learn to trust your intuition. More often than not it is right on. Remember research is multifaceted, and an *enjoyable extension* of your pastime, *not something that takes away from it.*

Research and reading could equal rewards.

When getting started, do not limit your research efforts to one particular area of interest. While you may be primarily a coinhunter, keep track of all treasure leads you happen upon. Record any and all information, interesting websites, newspaper articles, rumors, suggestions or even legends that come your way. Whether they pertain to a cache, a lost treasure, an old mine or a battleground, write them down. Your forte may be coin hunting, but keeping all your options open is wise. Who knows what will develop later as you continue your reading and documentation.

Record Keeping

As you begin your research, design a program or system that will allow you to easily and efficiently record the information you find. You might utilize a 3-ring notebook, files in your computer, or a 3x5 card file or a program on your computer. The choice is yours. When I started years ago 3x5 cards worked well, but now that I have a computer I keep my data stored in a word processing program, with folders such as *coinhunting*, *relics* and *caches*. I have sub files or folders as well with names like *schools, parks, homesites, churches, groves* and *carnival ground, landowners, access forms, websites, library contacts and maps.*

Whatever system you use, record *where* you came upon the information (library, historical society, personal acquaintance, etc.) .....the *book and page number* (if from reading), and the *date* you happened upon it. As you continue to pursue your hobby your files will slowly but surely grow. They may fill up fast, or at times seem to become stagnant...this is to be expected. Their growth is dependent upon the effort you put forth, and the resources available to you. If you persevere you will come up with the pieces to many of your puzzles.

Verbal Researching

One of the most important aspects of researching is what I call VERBAL RESEARCH - best of all it's easy and frequently produces instant results. What is verbal research? Simply - **ask everyone you come into contact with to help you find treasure!** Advertise your interest in the pastime, and ask everyone you know to help you find new treasures. Each and every individual you come into contact with can help - neighbors, co-workers, friends, relatives, business acquaintances and even strangers! Think about it! They all lead different lives, frequently in different towns and neighborhoods, have diversified contacts and interests, and each is a source of information waiting to be tapped. Multiply them by *their* friends, relatives, neighbors and co-workers, and you will understand how great the potential is!

Ask them if they know of old schools, churches, parks and homesites. Do they know of an old picnic area, or carnival grounds? If they draw a blank and do not respond immediately, ask them to think about it, ask their friends and relatives, and then check with them at a later date. Be persistent (not annoying), and the law of averages will be in your favor.

When verbally researching, make a special effort to ask the *senior citizen.* Senior citizens not only remember a great deal, they actually participated in events that will be of interest to you. They've witnessed many changes over the years, and I am willing to bet they will be happy to share *"the good ole days"* with you. I've found them knowledgeable, fascinating and often curious about metal detecting. They enjoy being able to help, and oh yes, don't be surprised however if you wind up with a new detecting buddy!

I can't emphasize enough that verbally seeking assistance really works! Ask **everyone** to help! You'll be amazed at the response you receive. If you are honest and direct in your questioning, you will receive honest and direct answers, and quite often immediate clues to better treasure hunting sites.

Research & The World-Wide-Web

No other medium in the history of the world has ever come close to matching the phenomenal impact that the computer has had on society as a whole. It has changed the way we work, live and even the way we think. It's capabilities are endless, and the amazing part is the technology is growing and growing. It's almost scary to think what the world will be like ten years from today!

Of course, not every home in America has a computer, but most of us have access to one. Whether at home, work, school or the local library, we can find one to use. As such we also have access to the Internet, and as a result the World-Wide-Web. What is the Internet? The Internet is a global network of computers, connected through a system of modems, satellites and telephone wires. It was originally designed as to preclude or to head off a world war.

The Web? A multifaceted network of sites available on the Internet that offers information, data, statistics and most important, research materials to you the detectorist/treasure hunter. If you can access the Internet, you can access the Web, and in doing so you will discover unlimited sources to help you in your quest for treasure. If you can spell it - you can find it on the web!

Research & The Library

Another area that's crucial to your success is reading. While it is time consuming, it's the single most productive aspect of your research effort. If you are not an avid reader become one! Hopefully the goal of finding old and valuable coins will be the incentive you need. You will not only come up with leads, new sites and finds, but will also become interested in your local history. It won't be intentional, *it will simply happen!* Never consider reading boring. Reading is easy, relaxing, educational and something that adds to your enjoyment of the hobby. If it rains on your day off and you are not able to get out in the field, do the next best thing and do some investigative reading. It could make your next trip out a fantastic one!

When reading about your community, think of yourself as an analyst. Your treasure hunting success depends upon your ability to uncover information from the past, and put it to work for you today. If you are after old and valuable coins, you will be seeking areas where people congregated at the turn of the century. If you are looking for that buried cache of money you've heard about, you will need to know the details of the story or legend that it surrounds. Like a detective, you will be piecing together bits of information, looking for clues, hoping to solve the crime, or in your case find the location of the treasure.

There are numerous libraries that can assist you. Local, county, and state libraries often duplicate materials, but each offers different documents that might add one more piece to a puzzle. Make up your mind to visit all of them at one time or another over the months ahead.

A great deal of your time at the library will be spent in the *"reference only"* area because much of the material you will be concerned with will be one of a kind - rare and impossible to replace. Take a pad, pencil and schedule your visits when you have time to spare.

Where To Start

Start with local history - your town, your city or county. Don't worry about anything on a larger scale for now. Keep your initial effort close to home. You will find that there's more to your local history than you imagined, and a great deal of useful information to record, and begin working with!

Look for materials that concern special events or celebrations that have taken place in surrounding communities. Books or pamphlets on centennial, bi-centennial and sesqui-centennial celebrations are ideal. They almost always contain *"fond looks back"* by older people in the community detailing their many experiences while growing up in the area. Such as where they played as youngsters, attended school, played baseball, football, and what special events they attended. If you come upon this type of material, be sure to jot down the name of the individual doing the reminiscing. He or she may still be living in the area, and a phone call or personal visit is recommended.

Remember your goals. Be alert for **any** mention of areas or sites that offer profitable detecting - where early residents of the community swam, picnicked, played ball, held carnivals, and church suppers. Also keep track of road and street names, or any label that was used to describe an area or locale.

Library Computers

If your library has computerized listings of its materials you will most likely find categories as **Author Search, Subject Search, Title Search, AND Alphabetical Search.** Under Subject Search enter the name of your community and county. Scan through the available materials, and if additional key words are needed try schools, parks, churches, celebrations, ghost towns, history, settlers, sports, recreation and legends. Take the same approach with the alphabetical listing, expanding your key words with treasure, crime, battles, railroads, markets, industry, maps, lakes and disasters. While scanning the screen, jot down any titles that might be useful. You should come up with a wealth of material to begin your research.

Microfilm

Inquire as to whether or not your library has a microfilm department. If so, find out what is available to you. Old area newspapers are often available on microfilm, and are extremely valuable research aids. Older newspapers not only describe various events that took place years ago, but offer photos as well. Be especially sure to read editions dated a day or two after a major holiday (the 4th of July, Memorial Day, Labor Day, etc.), for mention of how the community celebrated, and more important **where** they celebrated!

When reading older newspapers never overlook any section, column or department. Whatever the topic - crime, business, travel, sports or recreation, read it. Even the old *"chit chat"* or social columns can offer clues. Mentions of get-togethers or gatherings of any type can lead to treasure. Believe me, some of the most unlikely stories and topics have added some nice finds to my collection.

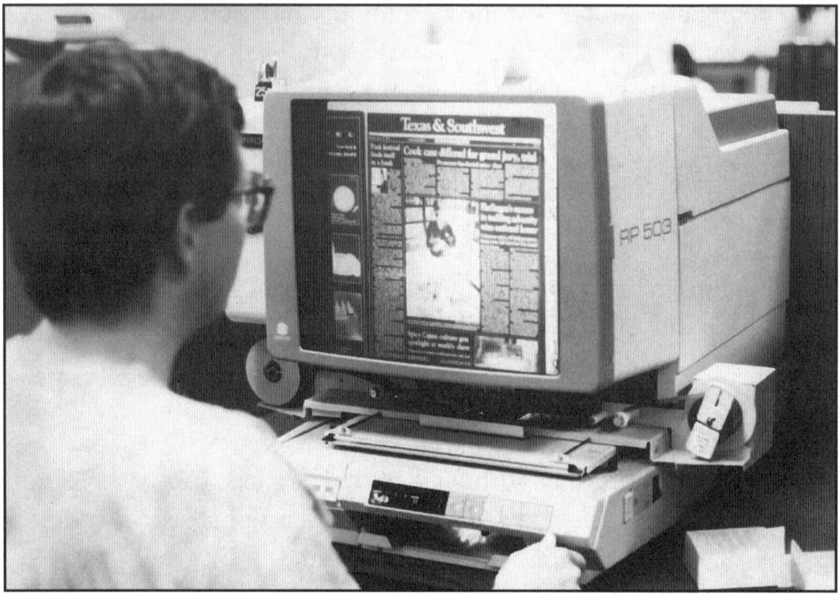

Be sure to peruse the older community newspapers on microfilm at the library.

Map Study

Last, but not least, ask what is available in the way of old maps or plattes of your area. Most libraries, no matter how small, will have something. Older maps will not be as precise or as detailed as those of today. With a little effort you can relate much of the information to the more recent ones. When possible, make photocopies so that you can study them at home.

Very early maps will frequently have notations of landowners. As you continue to learn more about your area, these names and notations will be very important. For example, when reading a local history book you may find references to locations like Johnson's Grove, Perry's Lake or Smith's field. By referencing early maps, you can often locate these sites. Naming community gathering places for the landowner was fairly common years ago, but these designations were often changed. So save those maps no matter how old or undetailed they may seem!

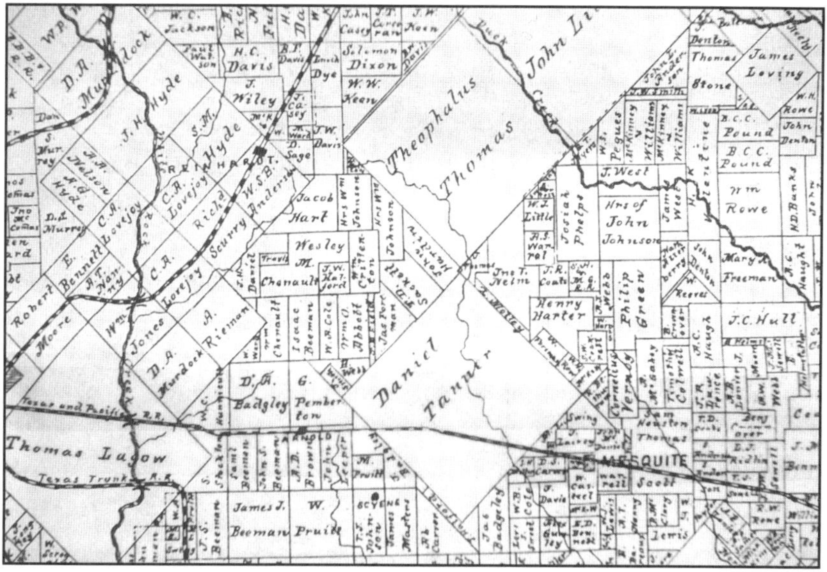

Very early maps were roughly drawn and usually indicated land ownership. Information that could prove very useful as you continue to research.

Map information is also available at the following websites:

www.carto.com/Older, rare maps
www.marketmaps.comCensus, zip code, data and state maps
www.kickapoo.com Historical maps prior to 1900
www.mapappeal.com Satellite photo maps
www.mapsonus.comService for finding specific address/locations
www.maptrade.org/Excellent for finding map dealers in your area
www.omnimap.com/Maps and geology supplies
www.ravenmaps.com Contour maps/gift maps
www.mapblast.com/Map locator service
www.mapmagic.comVarious map formats available

Historical Societies

Most towns, communities and counties have historical groups or societies whose purpose is to research and preserve the area's history. Find out about those in your area, and consider becoming a member. Dues are minimal, if not free. By joining you can do your part in helping to preserve the community's history, and have access to materials that may not be available anywhere else. Many old and rare documents are bequeathed to historical groups by individuals and families knowing that they will be used, preserved and appreciated.

As you might expect, most historical society members are long time residents of the community, and having the opportunity to be around them will also be to your advantage. You might consider donating a few of your more historic finds to such groups. They will be appreciated, perhaps displayed and most important be available for viewing by generations to come.

Topographical Maps

While all regional maps are important to your effort, one in particular stands out above the rest, and this is the **Topographical Map.** Topographical maps are often instant sources for new and productive detecting sites, and never become outdated. The dictionary defines topography as *"the science of representing the surface features of a region on maps and charts"*. While your treasures may be beneath the surface, the features on top often hold the clues necessary to find them. "Topo" maps, as they frequently are called, are precise, detailed and should be added to your arsenal as soon as possible. They can be purchased from map dealers in your area, but may also be ordered from the following government agencies:

If You Live East of the Mississippi:
U.S. Department of the Interior, U.S. Geological Survey,
907 National Center, Reston, Virginia 22092

If You Live West of the Mississippi:
U.S. Department of the Interior, U.S. Geological Survey,
Box 25286, Federal Center, Denver, Colorado 80115

When writing these agencies, indicate the state and county you live in, and request all materials relating to them. Topo maps are available in various scales and sizes, but the **7.5 minute series is best for your use**. The scale is one inch to 2000 feet, and the area covered is from 49 to 71 square miles, offering great detail, and a real sense of the area. They are in color with landmark indications for schools, churches, campgrounds, race tracks, railroads, mines and many other valuable sites.

For those of you who have computers be sure to write to:
MAPTECH
655 Portsmouth Avenue
Greenland, New Hampshire 03840
Phone: 603-433-8500/Email: www.maptech.com

This company offers topos on computer disk, making access extremely easy and quick, and you will like the many choices you have in accessing each and every area. You can narrow your search patterns, and zoom in on those that interest you most.

After you purchase topo maps of your area(s) make an effort to come up with **earlier ones**. A few are available from map dealers, and are reasonable in price. They might be photocopies, and not quite as detailed as the more recent one but this is not a problem. Major roads and streams will help you match old with the new, and allow for easy cross reference.

You will be amazed at how easy it is to find clues by comparing old and new topographical maps. You will quickly become aware of landmarks on the earlier map that are not present on the most recent, indicating perhaps a homesite that has burned or disappeared, an old

Topographical maps are extremely useful in the field to locate productive sites.

one room school that was torn down, or maybe a pond area that has been filled in. You might also notice that the old one room school house is now a residence, or that the old scout camp is now just a field or pasture.

Study topo maps thoroughly, and keep them handy. You will find yourself referring to them often. I've looked at mine for many years, and still come up with ideas and clues today.

The U.S. Geological Survey will also forward a listing of map dealers in your area. If possible visit their place of business or write them. They can often recommend other maps and materials that might help in your effort to find new treasure hunting sites you are not aware of.

Other government agencies that can assist you with your effort:

United States Printing Office
Superintendent of Documents,
Washington, D.C. 20402
(Maps, Atlases and other similiar materials)

Library of Congress
ATTN: Geographic and Map Division,
Washington, D.C.20540
(Historical Maps, Microfilm and Prints)

Websites that can be of assistance when researching:

www.loc.gov/
Library of Congress/Documents/photos/movies/and much more
www.carl.org/carl.html
Carl Corporation/Public access to large library database Excellent!

www.oclc.org:2000
OCLC online computerized library/Information research online

www.gopher.com
Gopher/Internet Directory and search engine

www.sunsite.unc.edu/reference
University of North Carolina/Davis Library reference info. Follow subcategories for excellent help.

www.lib.washington.edu/-
University of Washington Library services.

A few silver coins, a ring or two, and the afternoon was a success.

Save your clad coin finds and invest in a gold coin or rare coin that will increase in value as the years go on.

Chapter V - Coinhunting

Finding older, more valuable coins with a metal detector does not necessarily take a lot of skill. Finding them consistently does. I want to share a few ideas, theories and experiences with you so that your coin collection will not only grow in number, but will increase in value as well. Remember that whatever success I've had over the years is the result of common sense, dedicated research, and much trial and error.

Coin Values

If you are not a coin collector or familiar with coin values, you soon will be. In the meantime, purchase the latest *"Guide Book of United States Coins"* by R.S. Yeoman. Often called the Red Book, this annual publication will help you identify the coins you find, and tell you their worth.

When you start detecting, you will likely find newer clad coins which generally are only worth face value. However, you may find an older coin or two that has extra value, and it's important you know this. The Red Book is easy to use, and will assist you in knowing what to look for each time you recover a coin.

I also recommend recording your finds in a notebook no matter how insignificant they may seem. Write down where you searched, the number of coins you found, the denominations, and dates. It will only take a minute or two, and the documentation will be of great help later on. I've kept a log for the past twenty years, and the information often proves helpful when analyzing a particular site or area, or where to detect when I have limited time.

Develop A Positive Attitude

Have you ever had one of those days when you woke in the morning, felt well rested and raring to go? If so, then you know this *"positive"* feeling usually results in a productive and fulfilling day.

SITE ____________________

DATE	CONDITIONS	NOTABLE FINDS	DETECTOR USED

The above format is one I've found useful for recording coin finds from various sites.

You can achieve the same results if you learn how to obtain that same positive attitude when you go detecting. The power of positive thinking is an amazing thing, and can mean the difference in success and failure. Add little research and detailed preparation, and you have a winner. On the other hand if something is bothering you, or if you have put off doing an important task to *"escape"* with your detector and have some fun, you'll probably not be rewarded and rarely will you enjoy yourself. Take care of any obligations you have, and enjoy your treasure hunting a great deal more!!

Site Data Form

In order to get a feel for the many sites you will be hunting, come up with some method to compile data each time you venture out. Aside from the name and location of the site, notate the types of coins found, dates (if old and/or significant), detector used, unusual control settings and weather and ground conditions. You might also want to list any particular areas within the site that seem to produce more than others. After a period of time, you will be able to gauge a sites' productiveness, age and just maybe the best times to hunt it. It will also prove useful when you have a limited amount of time to hunt, and you want to make the most of it.

Go For The "Old"

Anyone with a metal detector can find coins. Go to any school yard in your neighborhood and you'll come home with many of them. All new and most likely tarnished. Yes it's spendable, but it's not collectible and it's not making the best use of your time. To find the older more valuable coins you must think "**old**". Searching recent sites will result in recent coinage, but those used years ago will provide you with vintage finds and make searching more fun. Ask yourself whether or not you'd rather find 50 clad coins in one hour or three or four silver pieces. I know what my response would be!

You might be saying, *"That's easier said than done. There aren't any old sites in my neighborhood"*. . Perhaps so, but what do you classify as your neighborhood? Your housing development, the adjacent five blocks, or your entire community? Expand your area of operation. Make an effort to go to the older sites, and I'll bet they're no more than fifteen minutes from your front door! If you've done your homework, researched, and talked to people you should already have quite a few to check out. If not, you're destined to be a schoolyard regular!

Your next response may be, *"Well I know of a site, but I have been told it's hunted out, and nothing of value is left to find"*.. My response to you is don't believe it! **No site is ever hunted out!** Others may have detected the area, but there's still treasure to be found, and more than likely the best the site has to offer. Why? Because no matter how slowly or meticulously you may hunt, it would take months if not years to exhaust a site. You would have to overlap coil sweeps in minute increments, maximize each and every function on your metal detector, and pick up every single piece of trash present to even come close, and frankly I enjoy these sites knowing that those who came before dug whatever surface trash was present. All I need to do then is listen closely to those subtle, barely audible, very deep signals. You know, those old turn of the century silver and gold coins.

TIP: Save your routine finds (clad coins) and invest in a gold coin or key coin that will increase in value as the years go on.

Increasing Your Odds

Certainly any of the sites I listed earlier will provide you with decent detecting, but choose those sites that offer the best return for your time. Consider the age of the area, the number of people who congregated there, and how frequently they made use of the site. In other words was the area used in the mid 1800's, or the 1940's? When the area was being used, were hundreds of people there at one time, or were there only a few? Was the site used every day of the

A good rule of thumb - older sites produce older coins.

week, only weekends, or at set times during the year? The questions you must ask are many, but in answering them you should be able to prioritize and make the best use of your time.

If you have more than one site in mind, and cannot decide which you should tackle first, always choose a **money site**. To determine if it is a money site, ask yourself the following question. *"Did people need money to access the area or to spend while there?"* For example, those attending athletic events usually had to pay admission. Those going to a carnival needed money for food, rides and souvenirs. These are money sites and the bottom line? **The more money that was used and exchanged, the more it was lost!**

Finding Coins In The City

If you live in a metropolitan area you may feel somewhat at a disadvantage because of the many buildings and paved areas, but it is not as detrimental to your metal detecting as you may think. Larger cities and communities have more people, and more potential for lost valuables and coins. Likewise larger cities are usually older increasing your odds for vintage type coins and treasures. Yes, condominiums and vast areas of concrete are a hindrance, but there are still plenty of areas that can provide excellent coinhunting.

First consider the schools. Very few new ones have been built in inner cities, and most of those present should be quite old (*newer consolidated schools are usually built in suburban or rural areas*). Next research the city's parks. Here again, the older your town or city, the older the parks and the better your finds. (*Many New York City parks continue to produce coins from the 1800's, and all have been detected hundreds, if not thousands of times*).

City schools have always had sports programs. Find the playing fields, and while we're on the subject don't forget the vacant corner lot. They were often the *"neighborhood"* sports arena and stadium for lack of anything better! Appearances can be deceiving, and if you are ever in doubt, a search with your detector will provide you with the answers.

Another reminder from earlier in this text - don't forget the grassy strips between streets and sidewalks. I once hunted these strips in a five block area of Trenton, New Jersey, and found fifteen silver dimes, 4 silver quarters, one silver half, and a few Indian Head cents. When added to the handful of clad coins, a silver ring and a few old matchbox cars my time was well spent. I never ignored these areas again!

One of the most available profitable sites you have in the large city is the older home. Do a little research and learn where they are located. They should give themselves away. Look for large shade trees, weather worn clapboard structures, lawns indicative of heavy use, and sidewalks that are sometimes heaved from tree roots and time. Older homes have both front yards and back yards that can produce old and valuable coins. Once you have found one of these gems, knock on the door and ask permission and remember my motto. **Nothing ventured - nothing gained!!** Being turned down is

After I first detected a grassy sidewalk strip, I never ignored the area again.

merely status quo. If there are fifty homes on a block, chances are you will have an opportunity to hunt 25 of them, and don't be surprised if neighbors see you hunting and ask you to search their yards out of curiosity.

Coinhunting in a metropolitan area is not difficult at all **if** you do your homework. With little effort you will find more than your fair share. You are probably never more than fifteen or thirty minutes from the suburbs and outlying rural areas so complaining about a lack of available sites is not valid. Your success or lack of it falls squarely on your shoulders!

Finding Coins In The Country

Having grown up and spent most of my life in rural New Jersey, I will admit a preference for detecting country areas. The farmlands of Hunterdon and Somerset Counties held many treasures, were rich in history, and almost always provided me with excellent detecting. Treasure hunters living in the city might assume that detecting the countryside is not very productive, but nothing is further from the truth! For one thing the fewer the people, the fewer detectorists there are to compete with!

One important thing to remember when detecting rural areas is that your manner of thinking, and your plan of action will change dramatically. What was obvious in the inner city will not be in the country. An athletic field in the city is obvious. In the countryside it's often just an open or overgrown field. An old school in the city will probably still be in use. In the countryside it might be a residence or have been torn down completely hidden by overgrowth. If you picnic in the inner city you more than likely do so at the park. If you live in the country you pack a picnic basket and go to the grove.

You must also be a keen observer when you travel the rural areas. Many of the best sites are hidden from view, often camouflaged, and waiting for the trained eye of an enthusiastic treasure hunter. I was very successful finding productive sites that to most people were only fields or wooded lots. Initially, I too passed them

This rural country school is contrary to anything you might expect in the city.

by never guessing that one day they would give up old and valuable coins. I learned to identify them quickly, and developed the attitude that every area had potential until proven otherwise.

Why do I enjoy rural areas so much? (1) They are usually not manicured or well kept allowing for easy accesibility and target recovery. (2) Most have never been searched with a metal detector. (3) Permission to search is usually easy to obtain. (4) Most of the sites have remained the same in physical appearance never being filled in or altered. (5) Most rural sites are relatively clean of modern day trash, and (6) they offer serenity, beautiful scenery, peace and quiet!

Snipers & Deepseekers

To be a successful coinshooter, I recommend that you add two **accessory** coils to your arsenal. First, a small "sniper" coil, which can be from four to six inches in diameter. (You shouldn't have a

Rural treasure hunting offers serenity, beautiful scenery, peace and quiet.

hard time choosing the size since the manufacturer probably only offers one for your model.) These smaller coils will not provide the depth your normal coil does, but will allow you to hunt trashy areas more effectively because of their very narrow search matrix (pattern).

Next consider the larger, deeper-seeking coil in the 9 inch to 12 inch range. It will add inches to your depth capabilities, and prove very beneficial when hunting areas that are relatively clean of trash, or have been detected heavily. The extra inch or two in depth you gain will often give you the advantage you were looking for, and enable you to find those deeper coins just out of reach of the seven to eight inch searchcoil and the "average" competitor.

Plan Ahead

This is a subject I hate to mention, but I have been metal detecting with friends, and had to take time out of my searching to assist them because they hadn't planned ahead. The number one problem - **dead batteries!** Be sure to charge your nicads the night before any trip, and if you are using alkalines or carbon zinc batteries be sure they are fresh. Weak or bad batteries can ruin a day, waste time, gas and in turn aggravate others with you.

Secondly, if you are traveling any distance from home consider the possibility that your detector might fail. Not a pleasant thought, but always a possibility. If you are going out with a friend perhaps he or she has another detector to take along as a spare. If not try to borrow one from a friend or club member, or consider purchasing an inexpensive model for use as a backup.

Be sure you have recovery tools with you that will accommodate different types of soil. A probe will not be much help in rocky terrain, nor will a shovel suffice at the local park.

Be prepared. Ground conditions can sometimes change dramatically. Having the right recovery tool can save the day.

Remember . . . . .

The site you are hunting may not provide many signals, but those that it does can be old and valuable. That means once again you must **scan slowly, overlap your sweeps, and listen closely** for signals that may be nothing more than whispers. Minute variations in your threshold are often your only clues.

If the area you're searching is old enough to produce Indian Head cents, know that your ID display (if you have one) will not identify them in the penny range, but just below. I mention this because a few years ago a treasure hunting friend of mine and I were sure we had located an old picnic grove, and eagerly started searching. It was at the time we both just purchased our first target ID detectors, and we put our faith in the target readouts. We searched for about an hour, came up empty-handed and left assuming we were wrong about this site. A month or two later, I insisted we give the area another try. When we did we began to dig those signals that read lower on our meters. You guessed it. We began recovering Indian Head cents, Shield and V nickels. As a result we slowed down, and dug each and every signal. Over the next three or four months we recovered over three hundred old and valuable coins. Those that did read penny and above were all silver!

Deep Coin Strategies

No matter what anyone tells you there's no sure fire way to find old and valuable coins everytime, all the time, however let me tell you about a few tactics that should increase your odds. They're not new and exciting tricks of the trade, but procedures based on logic, trial and error, and more often than not ignored by the majority of hobbyists.

First, let's tackle a question which has been asked for years. Are the older, more valuable coins deeper in the ground? In my opinion **yes!** Are they so deep we cannot find them with the technology we now have? Some, but the majority are fringe targets barely out of reach of the detectorist using average control settings, but readily available to the detectorist who puts forth the extra effort.

What do I mean by extra effort? As a treasure hunter you will naturally have ups and downs, but the one constant you have complete control over is your ability to analyze, adapt, and use every weapon in your arsenal. It might be changing to your larger coil, searching in the all-metal mode, or increasing your sensitivity. This "extra" effort must be put forth to find the better coins. Routine detector settings, careless scanning, inattentiveness and haste will usually result in a poor day, or at best an average one.

One last word on factory preset indicators. They are on detectors so the beginner can start searching quickly under "average" conditions. In difficult or overworked areas they will not offer you any advantage over the competition (who is probably utilizing the same preset marks). Learn to **fine tune** your detector, maximize its potential and come up a winner! *Average* settings are for *average* conditions and will result in nothing more than an *average* day of routine finds.

Come Up With A Plan

If you've done your homework, and you feel certain that a site has the potential to produce old coins, you need to develop a plan of attack. In other words, how should you adjust your detector's settings, and search the area? To develop your plan you must first know what the site has to offer. Is it trashy? Is the soil heavily mineralized? Is it wet or dry? Is the ground granular or rocky? Are there tall weeds or grass? Dig each and every target and analyze your finds. Is there one particular trash item that is predominant or do you have to contend with different types? Is your detector operating smoothly or is it erratic? Is there a lot of mineralization present? Having the answers to these questions, knowing what you have to contend with, and what changes or adjustments have to be made will enable you to quickly develop your plan.

Deep coin strategies pay off in older more valuable coins.

First and foremost always try to increase your sensitivity as much as you can until your threshold becomes erratic. When it chatters turn it back "just" to where it stabilizes. Any increase at all, no matter how slight will get you deeper, and that should always be your primary goal. On the flip side, however, you may find you have to turn your sensitivity **down** because of trash or heavy mineralization. If this is the case, don't worry. Turn it down! Threshold stability is always more important! Hearing subtle audio changes is not possible when your detector is erratic!

After maximizing your detector's sensitivity, scan the area slowly and listen for minute changes in your audio threshold. If, after a short period, you do not have any success switch to your larger searchcoil. More than likely this deepseeking coil working along with the maximized sensitivity will do the trick, and put you down where you need to be - where the older coins are, and your competitors were too lazy to go. Remember, that even with these two tactics your signals may still be whispers, so concentrate.

Finally if maximizing your sensitivity and adding the larger coil do not produce the desired results go to your all-metal mode, and **if** you have the ability to **manually** ground balance your detector do so. By manually ground balancing you will precisely "fine-tune" your metal detector, and gain a slight increase in depth over the automatic method. Automatic ground balancing, while improving each year, is at best an "averaging" out of ground minerals.

Hunting in *All-Metal* can be somewhat bothersome because you will hear each target audibly in your headset. The advantage is that you will be utilizing the deepest possible mode your detector has to offer. That coupled with the larger coil and increased sensitivity should produce results if the site has anything to offer. If your detector has a target ID display, allow that to act as a secondary check or discrimination tool, but do not let it be the final word. Remember again that it will only ID effectively down to about seven inches. Anything beyond that may not be strong enough to activate it, and the best discriminator you have is your digging tool!

Overcoming Trash Problems

When faced with a site that is extremely trashy use your smaller coil, and don't be afraid to decrease your sensitivity. The small coil will get you "in-between the cracks" for those coins masked by trash. Lowering your sensitivity will smooth out your threshold enabling you to define every signal you hear.

Scanning with the smaller searchcoil, and effectively overlapping your sweeps is tedious and very demanding. You will need a great deal of patience, and the ability to move at a snail's pace to do it right. However, don't be surprised when you find a few old and valuable coins at somewhat shallow depths ignored by the impatient and "average" hobbyist.

There's really no secret to being a successful coinhunter. Analyze the site, come up with a plan, utilize your options, be patient and you will more often than not succeed where most fail.

Summary

Typically you will find two distinctly different problems when searching for older coins. **One**, you will not come up with any signals or targets, or **two**, the area is so trashy it's impossible to distinguish good from bad without using extreme discrimination. Two extremes, but not unusual. Seldom will you walk into an old area that has potential and come up with what you are after without some obstacle and without making adjustments.

Your choices for the first scenario are as I described earlier - (1) increase your sensitivity as much as possible without causing your threshold to become erratic, (2) use a larger coil, and (3) search in the all-metal mode (manually ground balancing if possible). If you are faced with the latter problem of extreme trash, use your small sniper coil, be prepared to search at a snail's pace, and if need be, lower your sensitivity.

More Tips

• Wet or damp ground will usually provide greater depth whereas dry, parched conditions will reduce it.

• Using decent headphones will bring out those whispers that represent your primary goal - old, deep and valuable coins.

• When a trash target is rejected the detector's internal circuitry may take a second or two to recover or adjust itself, and during this period may skip over a good target. If your detector has a recovery speed control, set it for **fast recovery** when hunting trashy areas.

• Be sure to search under bushes or shrubs. Many have grown over the years, covering lost valuables in the process.

• After recovering a target always scan the hole again. Coins are often lost in multiples, and there may be others present.

• Always be careful when recovering a target - careless and hasty digging can damage a potentially valuable coin or find.

• When you've found a productive site, be sure to check out adjacent or fringe areas for additional finds. Try to ascertain where access routes, parking areas, and supplemental use areas were located. Never assume you've found "the" hot spot.

Chapter VI - Other Areas Of Metal Detecting

While most of your time will be spent searching for coins and jewelry there are other areas of the hobby equally enjoyable, and as your confidence grows so will your desire to participate in them. Like coinshooting they don't require a great deal of skill, but as with any endeavor a little preparation helps. Let your metal detector dealer suggest a book or two on the following subjects when you are ready to give them a whirl.

Beach/Water Hunting

As I mentioned in Chapter III beach and shallow water hunting is more popular than ever. Whether it's the ocean or an inland lake, water is like a magnet attracting outdoor enthusiasts by the thousands, and in turn providing the detectorist with many treasures. The primary objective of those who hunt the water and the water's edge is to recover gold rings and valuable pieces of jewelry. Finding one ring can equal months of finding coins, and those that diligently work the beaches find many!

Hunting the beach does not differ a great deal from hunting inland areas, but if you decide to hunt **in** the water you'll need to purchase a long-handled scoop for recovering your finds. A good one may cost as much as one hundred dollars, but when you consider that one ring can more than pay for this investment it's a small price to pay.

Most all metal detectors are waterproof up to the control box, but when you are searching in shallow water exercise caution. Take steps to insure that salt water doesn't penetrate the detector housing. **Salt water is extremely corrosive, and can ruin your detector.** If this happens, quickly rinse your detector with fresh tap water and allow it to dry out. Sometimes this quick action will do the trick and save you a great deal of expense. I would strongly suggest that if you intend to water hunt, invest in a waterproof submersible detector. All of the detector manufacturers offer such models, and while they are more expensive they excel in this environment and will offer you peace of mind! Otherwise all you need to find treasure **on the beach** is a sandscoop which can be purchased at your local detector shop. Sandscoops are like sieves, allowing sand to filter, but not coins and other valuables.

Prospecting

One of the very first applications for metal detectors was searching for gold. Gold is and always has been the king of precious metals, and with the refinements of the modern metal detector prospecting is even more popular today. Most all detectors are capable of finding gold nuggets in the ground, but the White's GM/4B was created especially for this task. Without going into boring statistics and details, this detector has a refined and unique ground balancing circuit that enables it to adjust quickly and easily to changing ground conditions. This is crucial since gold is most often found in highly mineralized soil. I want to emphasize however that electronic prospecting (searching for gold with a metal detector) is hard work and very frustrating. Many prospectors spend their entire life searching for this elusive metal, and never strike it rich. If you decide to become active in this area contact your state bureau of mining for information about prospecting in your area, and read a book or two about electronic prospecting. (*Finding Gold Nuggets II*, by Jimmy "Sierra" Normandi, is one of the best!)

Relic Hunting

Webster's Dictionary defines a relic as *"an object surviving from the past"*, and if we take that literally, everything we find with our metal detector is a relic. It might be an old farm implement, a child's toy, a knife or firearm or a piece of barbed wire. All are relics of some sort, but the typical relic hunter is looking for artifacts from the Civil War.

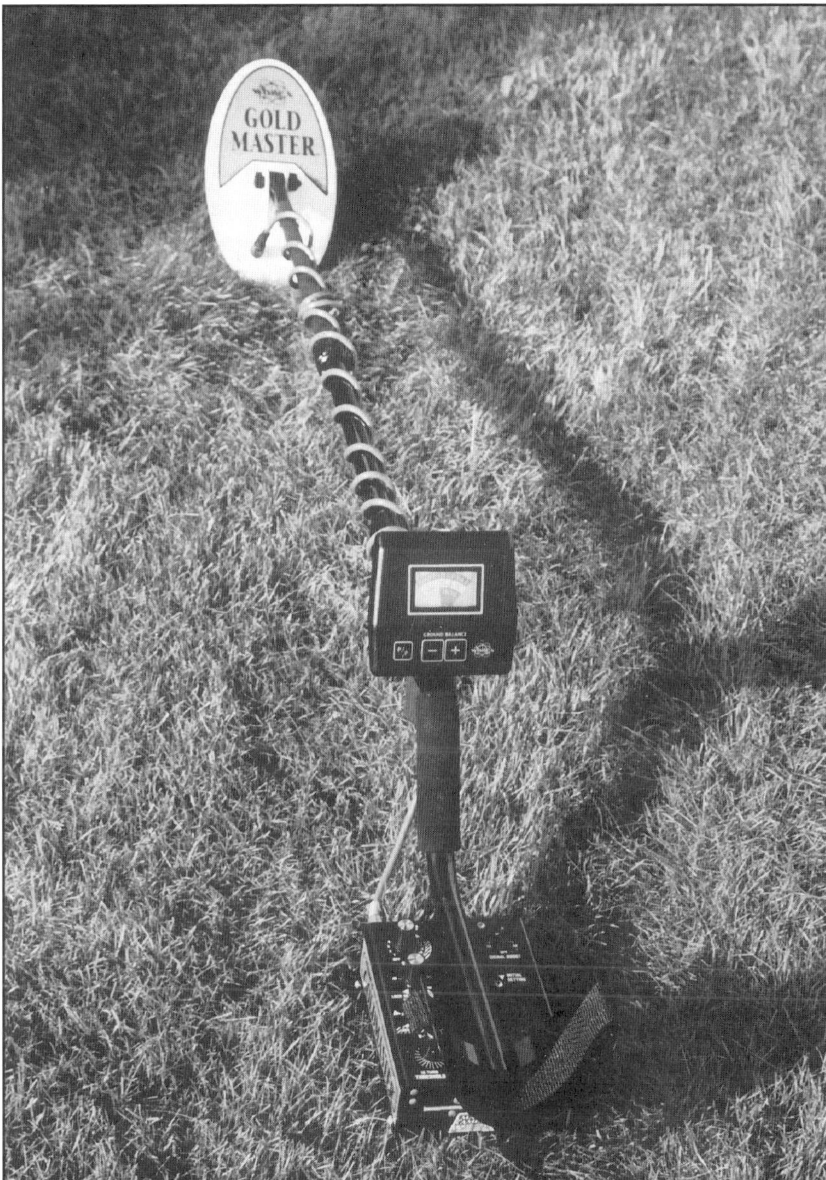

The White's Goldmaster 4/B - especially designed for electronic prospecting.

Relic hunters search for and find bullets, buttons, buckles, muskets, swords and related memorabilia. Many of these items are worth a great deal of money, but most relic hunters I know are reluctant to part with them, and usually keep them for their private collection or share them with universities and historical societies. *Many knowledgeable relic hunters have been responsible for finding sites unknown to historians.* If you intend to do any relic hunting dedicated research is a must! Knowing where various battles and skirmishes took place, as well as where the armies traveled, camped and trained is of great importance.

SPECIAL NOTE...

Metal detecting national battlefields, parks and historic sites is prohibited. Direct your relic hunting efforts to adjacent and nearby privately-owned areas, and if you are in doubt, ask first!! Likewise, get written permission to search on private property.

If you need more information on this exciting aspect of metal detecting pick up copies of *"Relic Hunter ..The Book"* and *"The World of the Relic Hunter"*, both written by Ed Fedory, and published by White's Electronics. Ed is without question today's best known authority on relic hunting with a metal detector.

Cache Hunting

Cache hunting is not an area where you start metal detecting and find a large amount of money. It takes research, perserverance, some intuition and a lot of luck! There are some who say that money caches do not exist in great abundance, but I beg to differ. I strongly suspect that they exist in greater numbers than any of us will ever know. Why? Because most were never documented in any fashion, the hiding place never shared, and those treasure hunters who have found them did not publicize their success.

The reasons people had for burying or secreting their money vary. Many just did not trust banks, and those that experienced the depression did not want to fall prey to another one. Whatever the reasons, a place was needed to secrete their money and many opted to

Inside and outside - always keep the possibility of a cache in mind.

bury it. Quite often these individuals were the only ones who knew where their money was hidden, not revealing its whereabouts to spouses or other family members. When sudden or accidental death claimed them, many families were unable to find the hiding place.

There are many tales about outlaw robberies and pirate treasure, but the typical money cache is the one the farmer, settler or eccentric hid and kept secret while going to his grave. I wish I could tell you I have found one or two, but for now they have eluded me. I have detailed leads on a couple, and one in particular I have searched for many times. I know the family in question well, believe their story, and have worked long and hard at finding it but so far nothing! Perhaps some day . . . .?

Chapter VII - Miscellany

Having covered the most important aspects of being a successful hobbyist, let me briefly address a few other items and areas that you might have questions about..

Two Box Units Or Cache Locators

Deepseeking two box detectors or accessory coils are designed for those instances where **extreme** depths are needed, and where the sought after object is quite **large**. They will detect larger targets to depths of 20 feet, but ignore anything small within the first two feet of the surface. This makes them ideal for large treasures that are buried deeply. What you must know is that you will not be able to discriminate while using them, and there are no guarantees that after digging down fifteen feet you won't find a large piece of trash. These deepseeking accessory coils are not for the average hobbyist, and used very seldom by those who do add them to their arsenal.

I know many detectorists who purchased two box units in order to find money caches confident that they were buried at extreme depths. Only one found what he was after, and that treasure was found at a depth of 18 inches! It's only my opinion, but if you were burying your money would you dig a hole six feet deep, or would you bury it so that it was quickly accessible? I think the answer is pretty obvious!!

Elliptical Coils

Elliptical coils are oval in shape, and designed for prospecting where they are better able to get in between rocks and crevices. A few manufacturers are offering them with other than prospecting machines, claiming that they work well in high trash areas for coins. I have not found this to be the case, and prefer the smaller coil in these instances.

Coil Covers

Coil covers are very thin plastic shields that fit on the bottom of your searchcoil to protect it from abrasions and accidental scuffing. They are not costly, worth having, and do not effect your detector's depth capabilities in the least! Should you become nugget hunter or prospector, I highly recommend their use because of the type of terrain you will be searching.

Recharhgeable Batteries

Most manufacturers offer rechargeable battery packs with their detectors, but if this was not the case when you purchased yours, consider buying one. Rechargeable or nicad batteries, as they are sometimes called, can be recharged up to 1000 times, saving you a great deal of money over the long haul. I use nicads as my main power source, but always have alkalines with me for added insurance. The only problem I have with nicad batteries is that they will discharge without any advance notice, whereas alkalines weaken gradually, and give notice that they should be replaced soon.

Today's rechargeable batteries are said to be free of memory characteristics (which develop through short booster charges). We will know this for a fact as time goes on. In the meantime, it's a good idea to allow nicads to discharge completely before charging.

Detector Cases And Bags

If you want to continue your hobby without interruption, keep your detector clean and do not abuse it. With a lot of common sense and a little maintenance it will last you a lifetime. However, if you are one of those individuals who tends to be careless, purchase a carrying case or bag to transport it. This will help protect it from knocks and drops when traveling, and when the season is over act as a storage container.

You'll certainly find many other accessory items available for your metal detector and the pastime in general. Most are nice to have, but not necessary for you to be successful in the field. Like anything else you enjoy, you will invest in a few of them anyway, and adding extra refinements to any enjoyable pastime is part of the fun.

Join A Club

There are thousands of metal detector hobbyists in the United States, and most belong to local treasure hunting clubs. Joining one of these clubs in your area will be of great benefit to you. These clubs usually meet once a month, dues are very minimal and the comraderie is fantastic. Each monthly meeting usually offers an informative program, a membership "finds of the month" contest but best of all a chance to meet others who share the same interests. You will be able to see what others are finding, what detectors they are using and what techniques they employ - all of which will make you a better treasure hunter. You may also find a treasure hunting partner to detect with.

Your local dealer will surely know of a club or two in your area, but if not write to **THE FEDERATION OF METAL DETECTOR CLUBS, INC.** This **national** organization is the spokes group for the pastime, and represents almost 125 treasure hunting clubs across the country. They routinely forward names and addresses of their member clubs to interested hobbyists, and will gladly assist you in finding a club in your area. Contact them at:

Federation of Metal Detector & Archeological Clubs, Inc.
5043 Strasbourg Way
Sacramento, CA 95842

or visit them on the internet at

FMDAC.COM

Club Hunts

Many treasure clubs hold hunts open to the public. If you participate in one of these events you will be competing for buried coins, and tokens redeemable for prizes. Treasure hunts are often all day events, where you'll meet other hobbyists, share experiences and most important have a lot of fun.

Always enjoy hunts for what they are, and never judge them by how you fare *(for there to be winners there must be losers)*. Every club that sponsors a hunt spends a great deal of time, money and effort to make it the very best. When the day is over be sure to thank them, and reflect on the fellowship and comraderie that took place. That's what these events are all about!

Metal Detecting Overseas

Whenever you have the opportunity to travel don't forget your metal detector. While the laws vary in every country, most allow detecting. As the first president of the Federation of Metal Detector and Archeological Clubs, I was privileged to travel and meet TH'ers from England, France, Spain and Japan. My most memorable experience was in Great Britain where I spent a great deal of time searching the countryside for Roman coins and relics many of which dated back almost 2000 years.

These bronze and silver pieces from early Roman times varied in appearance and size; some only as big as a nail head. Many were only in existence for very brief periods depending upon the era or time in history. Some were valuable, many were not, but they are all priceless to me, and being able to search this historic and beautiful country was something I'll never forget.

I would most definitely encourage you to visit the European countries for a little detecting. If you do, write the FMDAC for names and addresses of hobbyists in the various countries you will be visiting. They will be anxious to meet you, take you to productive areas, and in many cases offer lodging. It will be an experience you will enjoy and remember!

Roman coins found near Bath, England.

Be An Ambassador

No matter where your metal detecting takes you always remember you are an ambassador for the hobby. What you say and how you act reflects on all of the many thousands who participate. If we are to continue to enjoy the pastime for years to come it is imperative that we honor the traditional treasure hunting code of ethics.

1. Always Respect The Rights and Property of Others
2. Observe ALL Laws
3. Aid Law Enforcement Officials Whenever Possible
4. Never Destroy Priceless, Historical or Archeological Treasures
5. Leave the Land and Vegetation As It Was
6. Remove All Trash and Litter When You Leave
7. Remember All Treasure Hunters will be Judged by the example **you** set.

Always conduct yourself in a courteous and considerate manner.

Conclusion

I sincerely hope that this text has shed some light on the fascinating hobby of metal detecting. Once again it's a pastime that will provide you with outdoor exercise, a lot of excitement and a profit to boot! You may not become a millionaire, but you will certainly be richer in many ways.

I hope someday the metal detecting hobby will be as popular to the masses as golf, hunting and fishing - time will tell. For now I urge you to get involved and reap the rewards. You will have a head start in one great fun-filled pastime! Remember, no matter what you find with your metal detector *the real treasures will be the friends you make, and in the search itself.*

Good hunting!

Glossary Of Metal Detecting Terms

AC (Alternating current) - Metal detector mode that requires searchcoil motion to respond to metal.

AIR TEST - Testing the depth capability of a metal detector by moving various metallic samples under it's searchcoil. Air tests are important, but not really an accurate barometer of how deep a detector will detect in the ground.

ALL-METAL - A mode that allows for acceptance of all metals. Frequently used in prospecting and when searching for the deeper targets.

AUDIO-IDENTIFICATION - Circuitry that produces different tonal responses for various metallic targets, and based on their conductivity.

AUTOTRAC (Automatic Ground Tracking) - A feature that continually and automatically readjusts the ground balance (ground rejection) of a metal detector during searching.

AUTOMATIC TUNING - Circuitry that continually readjusts audio threshold to it's preset level after drift and target rejection.

AUTOMATIC GROUND TRACKING - A feature on some metal detectors that continually adjusts ground balance as the ground conditions change.

BACK READING - A false signal caused by trash items coming into contact with the bottom of the searchcoil.

BLACK SAND - One of the most extreme components of non-conductive negative ground mineralization. Also called magnetite, iron oxide and magnetic sand.

CACHE - A secreted or lost hoard of valuables.

CO-AXIAL COIL - A searchcoil that has one transmit winding in between two receive windings, identical in diameter and aligned on center.

COIN DEPTH INDICATOR - A visual readout of a target's depth. Usually calibrated for coin sized objects, and may fluctuate with the use of smaller or larger coils.

CONCENTRIC COIL - A searchcoil using one or more transmit windings, and one receive winding. They are unequal in diameter and aligned on a common center.

CONDUCTIVITY - The measurement of eddy currents on a metal target's surface.

CONTROL HOUSING - That part of a metal detector that encases the circuitry, controls and power supply.

CO-PLANAR COIL - A searchcoil design in which the transmit and receive windings are in the same plane.

2D COIL - A Co-Planar searchcoil with two "D" shaped transmit and receive windings, back to back, but overlapping.

CUSTOM PROGRAM - Preferred choices or settings on a computerized metal detector that are saved or stored for future use.

DC (Direct Current) - Used to refer to metal detector modes that do not require coil motion to respond to metal targets.

DEPTH PENETRATION - The ability of a metal detector to send a signal into the ground and produce a metallic target signal.

DETECTION PATTERN - That area of the searchcoil's penetration where metallic targets can be located. It's conical or oval in shape and changes in size according to the target's surface.

DETUNING - The process of lowering the audio threshold into the null area, enabling the detectorist to narrow the signal width for better pinpointing.

DISCRIMINATION - Circuitry that ignores responses from specific metallic targets, while allowing others to respond audibly. Of great use in eliminating unwanted trash items.

DOUBLE BLIP - A response or signal that usually indicates an elongated object such as a nail.

DRIFT - A loss of audio threshold due to temperature changes, battery condition or mineralization in the soil.

EDDY CURRENTS - Small currents emitted from the surface of detected metals which produce an electromagnetic field, received by the searchcoil and audibly transmitted to the detectorist.

ELECTROMAGNETIC FIELD - The pattern extending from the top and bottom of a searchcoil. Created by the oscillator frequency current in the transmit winding of a searchcoil.

ELECTRONIC PINPOINTING - Automatic feature or mode on metal detectors that narrows target response for the purpose of pinpointing and recovery.

ELLIPTICAL COIL - A searchcoil with an ellipse or oval shape. Designed for detecting in nooks and crannies, and usually associated with prospecting.

FALSE SIGNAL - A misleading signal or response caused by overshoot (retuning), ground conditions or hot rocks.

FARADAY SHIELDING - Conductive wrapping of searchcoil windings to eliminate electrostatic interference.

FERROUS OXIDE - Decomposed particles of iron (mineralization)

FREQUENCY - The number of alternating current cycles produced by the transmit oscillator per second. Measured in Hertz (Hz), or cycles per second. Most detectors manufactured today operate in the Very Low Frequency range of 3 to 30 kHz.

FREQUENCY SHIFT - A feature or option on a metal detector that alleviates cross talk or interference between two detectors of the same frequency when working in close proximity.

GROUND BALANCE - The adjustment or operation of a metal detector to eliminate ground mineralization (most notably from iron mineralizaton).

HOT ROCK - A rock containing a higher concentration of ground minerals than the surrounding area to which the metal detector is balanced to. In the motion mode a hot rock will produce a positive response but produce a null in the all-metal mode.

INTERFERENCE - Hindrance of performance due to sources outside a metal detector causing static and unwanted false signals.

LCD (LIQUID CRYSTAL DISPLAY) - Used on metal detectors as a graphic or visual readout as opposed to a meter/needle indicator.

LOOP - A non-metal housing containing the transmit and receive windings used to generate and receive signals from metal targets. Also called Searchcoil.

MEMORY - A computerized metal detector's ability to retain operator selected settings for future use.

MENU - A display screen that allows the operator to choose among different features and programs.

METER - That part of a metal detector that provides visual information to the user. Meters can be either LCD or meter/needle and provide intensity of the signal, target depth, target identification, type of metal detected and battery condition.

MINERALIZED GROUND - Soil that contains conductive or non-conductive components.

MIXED MODE - A special metal detector mode that combines all-metal and discriminate features into one operating mode.

MODE - A particular circuitry or channel of operation on a metal detector (i.e. Discrimination Mode, All-Metal Mode, Pinpointing Mode, etc.).

MOTION DISCRIMINATOR - A detector requiring searchcoil motion to activate ground balance and discrimination.

MICROPROCESSOR - An electronic component or chip that can be programmed to perform certain tasks or functions. Many detector manufacturers are incorporating them into today's detectors.

NEGATIVE GROUND - Soil containing magnetic minerals which usually have a negative or nulling effect on a metal detector's threshold.

NEUTRAL GROUND - Soil containing no non-conductive or conductive properties.

NICAD (NICKEL CADMIUM) - Battery that can be re-charged repeatedly.

NON-FERROUS - Not of iron. Generally metals of the precious class (silver, gold, etc.), however can also refer to aluminum, lead, zinc and brass.

NON-MOTION - A mode of operation that does not require the searchcoil to be moving for detection to occur.

NOTCH DISCRIMINATION - Circuitry that allows a window of either acceptance or rejection within the discrimination range, and is uneffected by discrimination being used above or below that point. Typically used to accept the nickel range while rejecting the pull tab range. Can also be used to accept only one specific range of targets while rejecting others.

NULL - The area just below audio threshold. This is also what happens when a detector momentarily loses it's audio while rejecting a target.

OSCILLATOR - Electronic circuit designed to generate a rate of current frequency to power a searchcoil's transmit windings (producing an electromagnetic field).

OVERLAP - Refers to advancing the searchcoil in small increments to insure that targets will not be overlooked.

OVERLOAD - Occurs when the receiver of a metal detector becomes overwhelmed due to too much signal (ground and/or target).

OVERSHOOT - A false signal heard as a searchcoil passes over a rejected target when operating in the non-motion, discriminate mode. Can also refer to any delayed audio response, such as occurs when passing the coil over a hot rock while in the all metal/automatic threshold mode.

PHASE - The length of time between eddy current generation sustained on a metal's surface and the resultant secondary electromagnetic field effect on the searchcoil's receive winding. Relates to target conductivity.

PINPOINTING - Finding the detected target's exact location. Usually accomplished by switching to the pinpoint mode or by detuning.

PI (Pulse Induction) - A type of metal detector circuitry that operates differently than the standard Transmit Receive or Very Low Frequency instruments. Primarily recommended for better results in salt water.

POSITIVE GROUND - Soil containing conductive minerals that have a positive effect on a metal detector's threshold.

PRESET MARKINGS - Indicators or markings on a metal detector control panel to assist the user in setting controls.

PRESET PROGRAM - A selection of a computerized metal detector that automatically selects all the features suited to a particular type of searching condition.

PROSPECTNG - Searching for gold in its natural state (nuggets).

RECEIVE WINDINGS - Coiled wire inside the searchcoil designed to receive the secondary electromagnetic field emitted from a detected target.

REJECTION - Non-acceptance or cancellation of a target response by discrimination circuitry.

RELIC - An artifact of the past.

S.A.T. (Self Adjusting Thrshold) - Automatically resets the threshold to correct for any minor ground, temperature, or electrical changes that typically cause threshold variations.

TWO BOX - A radio frequency detector having transmit and receive windings separate from one another. These detectors are capable of detecting large objects at great depths, but will ignore smaller items.

SEARCHCOIL - Plastic housing (usually circular) containing the coil windings to transmit and receive signals from ground and targets. Configuration of these windings can vary.

SENSITIVITY - The capacity of a metal detector to note changes in conductivity. Generally speaking the more sensitivity used the greater the depth, however some soils may not allow for this.

SIGNAGRAPH - A graphic display of a pattern characterizing a target's electrical and/or magnetic properties.

SILENT SEARCH - Operating a metal detector below audio threshold. A few models only operate in this mode.

SKID PLATE - Protective cover for the searchcoil's bottom....helps protect from abrasions, wear and tear.

STABILITY - The ability of a metal detector to maintain smooth operation without interference or unstable behavior.

SURFACE BLANKING - A metal detector feature that ignores targets (good or bad) close to surface of ground, but accepts those of greater depth so long as they are not directly beneath those being blanked out.

SURFACE MOUNT - The process of mounting electronic components on the surface of a printed circuit board, rather than through it, allowing for more technology over a much smaller area. This creates components that are lighter, smaller and more portable.

TARGET MASKING - Occurs when large concentrations of trash overpower better targets causing them to be rejected rather than detected.

TEN TURN CONTROL - A control that can be manually rotated ten times to cover the full function of it's purpose. Ten turn controls are usually found on prospecting detectors and used for precise ground balancing.

TEST GARDEN - A mapped plot of buried targets at various depths to aid in learning the characteristics of a metal detector.

TH'er, TH'ing - Short for Treasure Hunter/Treasure Hunting.

THRESHOLD - Minimum audio sound when detecting. Generally the best way to operate for maximum performance.

TONE CONTROL - A control that allows for adjusting range of audio frequency or pitch.

TOUCH PAD - Switches encapsulated in plastic that allows the operator to select different functions or features.

TRANSMIT WINDING - Wire coil inside searchcoil designed to send out electromagnetic field into ground.

TR (TRANSMITTER/RECEIVER) - Generally a term used to refer to detectors manufactured a few years ago. It's important however to understand that most detectors available today are still TR units, but utilize VLF frequencies, and can better ignore ground mineralization.

TWO BOX - A radio frequency detector having transmit and receive windings separate from one another. These detectors are capable of detecting large objects at great depths, but will ignore smaller items.

VISUAL DISCRIMINATION - The ability of a metal detector to determine trash or non-trash by means of visual indications.

VISUAL DISCRIMINATION INDICATOR (VDI) - A visual indication as to the type of target a metal detector is detecting.

VLF (VERY LOW FREQUENCY) - Very low frequency (3 to 30 kHz).

VLF/DISC - Detectors capable of mineral free operation in both discriminate and all-metal modes.

VLF/TR - Detectors that operate in both All-Metal, Ground Balance Mode and Non-Motion Discriminate, Non-Ground Balance Mode.

VOLUME CONTROL - Control that sets loudness of target response. Should not be confused with threshold volume.

WIDE SCAN COIL - A searchcoil that can detect targets across it's full diameter.

ZERO DISCRIMINATION - Discrimination circuitry that allows for acceptance of all metals at zero setting.

Dick Stout - has been a treasure hunting hobbyist, detectorist for over 25 years and still enjoys the anticipation and thrill of the search.